T0291144

# Principles of Aeroelasticity

# Principles of Aeroelasticity

## Rama B. Bhat

**CRC Press**
Taylor & Francis Group
Boca Raton  London  New York

CRC Press is an imprint of the
Taylor & Francis Group, an **Informa** business

CRC Press
Taylor & Francis Group
6000 Broken Sound Parkway NW, Suite 300
Boca Raton, FL 33487-2742

© 2016 by Taylor & Francis Group, LLC
CRC Press is an imprint of Taylor & Francis Group, an Informa business

No claim to original U.S. Government works

Printed on acid-free paper
Version Date: 20160229

International Standard Book Number-13: 978-1-4987-2472-2 (Hardback)

**Visit the Taylor & Francis Web site at**
**http://www.taylorandfrancis.com**

**and the CRC Press Web site at**
**http://www.crcpress.com**

# Contents

# *Preface*

This book has evolved out of a one-semester course on principles of aero-elasticity given at Concordia University, Montreal. Even though the concept of aeroelasticity is quite important (mainly in the design of aerospace structures), there are no suitable textbooks in this area.

This book does not suppose that the student has already taken courses in the theory of vibrations. With the background of mechanics and strength of materials and solutions of differential equations, one should be able to follow the material in this book.

Aeroelasticity is a subject that deals with fluid–structure interaction problems in general, even though, traditionally, it evolved because of a special need in the design of aerospace structures. This book deals with some nonaerospace topics also.

There are excellent reference books for people engaged in research in this area, such as those by Bisplinghoff, Ashley, and Halfman (1955); Fung (1955); and Curtiss, Sacanlan, and Sisto (1989), edited by E. H. Dowell. This book is mainly intended for undergraduate students and first-year graduate students in order to introduce them to aeroelasticity and may give them an appreciation of this exciting area.

# *Author*

**Rama B. Bhat, PhD,** is a professor of Mechanical and Industrial Engineering at Concordia University. His research areas include dynamics of structures, vibration of continuous systems, rotor dynamics, sound transmission into aircraft fuselage, dynamics of microelectromechanical systems, ride dynamics and biodynamics.

He has published widely in these areas. He has taught both undergraduate and graduate courses in the above areas, and in particular, he has taught the course on Principles of Aeroelasticity for more than two decades. He received the NASA Award for his contribution as a member of the team that developed the "PROSSS-Programming Structured Synthesis System." He is a fellow of several technical societies such as ASME, CSME, and the Engineering Institute of Canada.

# 1

## *Introduction*

In studying the mechanics of systems, we always start with the equilibrium of objects under the action of external forces. When the system is in motion, we also employ Newton's laws of motion, which relate the motion with the cause of motion, which is the force. We start the study with simple systems such as particles, systems of particles, and rigid bodies, which are systems of particles held rigidly together. When we extend the study of dynamics of systems to include the elastic property of the systems also, we lump such elasticity in the form of idealized spring elements and analyze the repetitive motion of the system about its equilibrium position, which is vibration.

In reality, the mechanical systems that we come across, particularly in engineering applications, are not that simple. The mass is distributed over the entire system and so is the property of elasticity. These are termed as continuous systems. In addition to the mass and elasticity, the system also possesses a distributed damping property. In aerospace applications, the weight of the system is very crucial, and we always try to minimize the weight. Naturally, as we try to minimize the weight, the structure becomes more flexible and the strength considerations are extremely important.

When the external forces acting on the mechanical system are independent of the deformation in the system, we can employ the techniques in the classical theory of elasticity in order to study the deformation of the system under the action of the external forces. However, in aerospace applications, the aerodynamic loads acting on the structure depend on the structural deformations also. As an aircraft flies against the wind, the wind load on the aircraft wing will depend on the attitude angle of the wing heading against the wind. The wind load will deform the wing torsionally, and this increases the attitude angle with a consequent increase in the wind load. This situation may lead to an unstable situation leading to the failure of the wing to withstand the wind loads.

We have seen flags and boat sails fluttering in the wind. Tall palm trees sway back and forth in steady winds, as do chimney stacks. High-tension electrical transmission lines with very long spans oscillate slowly in steady winds and start galloping if the wind velocity is quite large. Flexible pipes carrying fluids can start humming due to its oscillations even when the flow velocity is constant. Suspension bridges are subject to buffeting oscillations in steady winds. All of these fall into the same class of phenomena, which are aeroelastic in nature.

Aeroelasticity is the study of the effect of aerodynamic forces on elastic bodies. In the area of elasticity or vibrations, the loads are assumed to be unaffected by the deformation or motion of the structure. However, the aerodynamic forces depend critically on the attitude of the body relative to the flow. The elastic deformation plays an important role in determining the external loading itself.

Stability of a structure in wind is an important consideration in aeroelasticity. For a given configuration of the elastic body, the aerodynamic force increases rapidly with the wind speed; however, the elastic stiffness is independent of the wind. Hence, there may exist a critical wind speed at which the structure becomes unstable. Such instability may cause excessive deformations and may lead to the destruction of the structure. This is termed a divergence failure.

Flutter stability is another important consideration in aeroelasticity where the amplitude of time-dependent structural deformations under dynamic conditions continues increasing, leading to structural failure.

In the ensuing discussion on aeroelasticity, Chapter 1 provides some ideas on the elementary aerodynamics in order to get an understanding of the types of loads acting on the aircraft structure. Then a simple analysis to illustrate the phenomenon of divergence is carried out on a one degree of freedom model of an aircraft wing. A complete analysis of an aircraft structure must consider the fuselage, wings, tail structure, etc. and can only be carried out using a finite element analysis. However, our interest is limited to the study of the interaction of aerodynamic forces and the elastic structure, and this is predominantly seen in the wings. Hence, we consider the fuselage as a rigid support and consider the wings as cantilever appendages fixed to the fuselage. Since the divergence phenomenon is predominantly due to the torsional response of the wings, we illustrate this phenomenon on a single degree of freedom model of the wing where the torsional stiffness of the wing is idealized in the form of a torsional spring attached at the elastic axis of the wing. The effect of adding a control surface to the wing is illustrated on the same simple model.

Before starting to discuss flutter, fundamentals of vibration theory are introduced using single degree of freedom and two degrees of freedom systems. Subsequently, flutter phenomenon is discussed using a two degrees of freedom system model for the aircraft wing. Dynamic response of aircraft wings under wind loads, in general, is discussed after this.

An aircraft structure is a flexible, continuous system in reality. At this stage, the wing is treated as a continuous system and the motion is described by differential equations. Exact methods and approximate methods of analyzing these differential equations are discussed. Classical methods, Rayleigh-Ritz techniques, Galerkin's technique, influence coefficient method, and finite element methods are discussed.

Some nonairfoil types of aeroelastic problems are discussed in Chapter 10. Pipes with fluid flow and plates with fluid flow are some examples.

# 2

## Elementary Aerodynamics

## General Concepts

Whenever wind blows past a stationary obstacle, the obstacle is subjected to a force. A blowing wind is able to rustle the leaves on a tree, while a strong wind can even uproot a tree. Stationary structures such as chimney stacks, tall buildings, bridges, electrical transmission lines, and broadcast transmission towers are subjected to forces from blowing winds.

An object moving in atmosphere has to move against a stationary mass of air, which prevents it from moving. The moving object has to overcome this resistance. This resistance from the stationary mass of air is similar to the force exterted by blowing wind on stationary obstacles. Hence, objects falling through atmospheric air, automobiles traveling on highways, and airplanes moving through ambient masses of atmospheric air are all subject to wind loads.

In order to understand the types of loads acting on aircraft structures, we will have to understand the behavior of air flow over obstacles.

## Stagnation Point

Whenever air flows around a body, part of it goes to one side and the rest to the other side. The streamline that follows the flow division (that divides on the upstream side of the body and joins again on the downstream side) is called the dividing streamline. At the point of division, the velocity of air relative to the body is zero. This point is called the stagnation point.

## Ideal-Flow Theory

Ideal-flow theory stipulates that the fluid is incompressible and nonviscous. A differential equation fundamental to the ideal-flow theory is Laplace's equation. Many useful insights in aeroelasticity can be obtained using this theory. When speeds are small compared to the speed of sound in a fluid, the fluid can be considered incompressible. Except in a small layer (called "boundary layer") close to the solid, viscous effects are negligible and the flow there can be considered nonviscous.

## Rotation

Consider a tank of liquid rotated about its vertical axis. The element of fluid shown (Figure 2.1a) rotates as time passes from $t_1$ and $t_2$, but does not deform. This is a case of fluid rotation. Consider fluid flow between two horizontal plates (Figure 2.1b). The bottom plate is stationary and the top is moving to the right with velocity V.

The velocity distribution is linear. Element faces that were initially vertical at $t_1$ rotate clockwise and are slanted at $t_2$. Horizontal faces do not rotate. If the average rate of rotation of two initially mutually perpendicular lines is zero, the flow is irrotational. In this case, it is not so and hence flow between parallel boundaries is a case of rotational flow.

In the case of irrotational flow, we have

$$\frac{\partial w}{\partial y} = \frac{\partial v}{\partial z}; \ \frac{\partial u}{\partial z} = \frac{\partial w}{\partial x}; \ \frac{\partial v}{\partial x} = \frac{\partial u}{\partial y}$$

where u, v, and w are velocity components along x, y, and z directions, respectively.

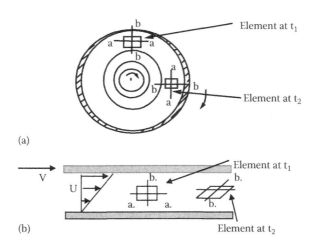

(a)

(b)

**FIGURE 2.1**
Tank of liquid rotated about vertical axis. (a) Liquid element rotating about a vertical axis and (b) liquid element between two horizontal plates.

## Vortices

A flow in which the streamlines form concentric circles is called a vortex. There are two types of vortices:

1. A forced vortex, where velocity increases linearly from the center of rotation
2. A free vortex, where the product of the velocity and the radial distance is a constant—that is, Vr = C.

## Separation

Flow patterns that are developed by ideal-flow theory are for irrotational flow. Flow of air or water in a region where the streamlines converge usually approximates irrotational flow closely. However, in regions where boundaries turn away from the flow so as to cause the streamlines to diverge, the flow usually "separates" from the boundary and a recirculation pattern is generated in the region. This phenomenon is called separation. Ideal flow (Figure 2.2a) experiencing no separation and real flow (Figure 2.2b) past a plate, which experiences separation, are shown in Figure 2.2.

## Drag and Lift

Consider the forces acting on the airfoil shown in Figure 2.3. Vectors normal to the surface represent normal pressure. Velocity of the flow over the top of

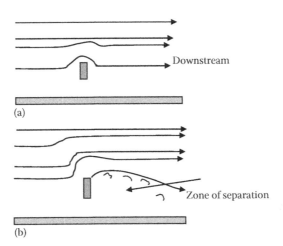

**FIGURE 2.2**
Figure flow separation. (a) Ideal flow experiencing no separation and (b) real flow experiencing separation.

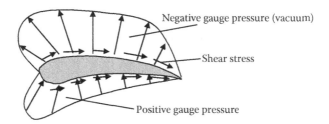

**FIGURE 2.3**
Forces acting on an airfoil.

the airfoil is greater than free stream velocity and hence the pressure over
the top is lower than the ambient/gauge pressure (resulting in a negative
pressure coefficient). Velocity along the underside is less than the magnitude of free stream velocity and hence the pressure there is greater than free
stream ambient/gauge pressure (resulting in a positive pressure coefficient).
A negative pressure coefficient over the top and positive pressure coefficient
along the bottom contribute to the lift. It should be noted that the absolute
pressure over the surfaces can never be negative.

Except on the front of the airfoil, shear stress acts essentially parallel to
the free stream direction, contributing largely to the drag of the airfoil. The
pressure and viscous forces acting on a differential area of the surface of the
airfoil are shown here. Pressure force $dF_p = p\,dA$ and viscous force $dF_v = \tau\,dA$
are shown (Figure 2.4). Taking components normal and parallel to the free
stream direction, we get the lift and drag forces, respectively, as

$$F_L = \int (-p\cos\theta - \tau\sin\theta)\,dA$$

$$F_D = \int (-p\sin\theta + \tau\cos\theta)\,dA$$

The coefficient of drag $C_D$ and coefficient of lift $C_L$ are defined as

$$C_D = \frac{F_D}{A_p\rho\left(V_0^2/2\right)}$$

$$C_L = \frac{F_L}{A_p\rho\left(V_0^2/2\right)}$$

Note that the nondimensional force will be derived later, in the form

$$\pi_4 = \frac{F}{l^2\rho U^2}$$

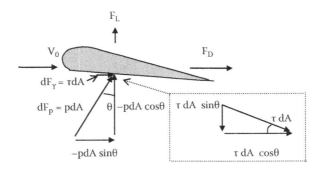

**FIGURE 2.4**
Forces acting on a differential area of an airfoil.

where $l^2$ denotes an area and U is velocity. In nondimensionalizing $F_L$ and $F_D$, we have used $A_p \rho \left( V_0^2 / 2 \right)$, where $\frac{1}{2} \rho V_0^2$ is called the dynamic pressure and $A_p$ is the projected area of the airfoil.

Approximate values of CD for various bodies are given in the following table (Shapiro, 1964):

| Shape | Drag Coefficient |
|---|---|
| Sphere | 0.47 |
| Cone (with apex facing flow) | 0.50 |
| Cube | 1.05 |
| Long cylinder | 0.82 |
| Short cylinder | 1.15 |
| Streamlined body (airfoil) | 0.04 |

Wind tunnel test results on a sedan are shown in Figure 2.5.

Centerline pressure distribution for a conventional sedan by a solid line and that for one with a rear deck lip causes pressure on the rear of the car (behind the lip) to increase, thereby reducing drag. It also decreases the lift, thereby improving traction. Of course, the lip itself produces some drag and there is an optimum lip height that will cause an overall drag reduction.

## Lift

It was explained earlier that the lift is generated by the pressure difference between top and bottom surfaces, caused by velocity differences on them. In

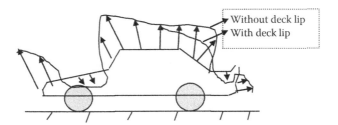

**FIGURE 2.5**
Effect of a rear deck lip on model surface pressure coefficients normal to the surface.

order to understand the cause for such velocity differences between top and bottom surfaces of airfoils, we must understand the phenomenon of "circulation," which is the basic cause of lift.

Consider flow along a closed path as shown in Figure 2.6. The stream velocity is V and the tangential component is $V_L$. Circulation is the quantity obtained by integrating the tangential velocity component around the body—that is,

$$\Gamma = \oint V_L \, dL$$

where $\Gamma$ is the circulation. The circulation for an irrotational vortex can be obtained as

$$\Gamma = \int_0^{2\pi} \frac{C}{r} r \, d\theta = \int_0^{2\pi} C \, d\theta = 2\pi C$$

where tangential velocity at any radius for the free vortex is given as $C/r$.

One way to physically induce circulation is to rotate a cylinder about its axis. The velocity of the fluid close to the cylinder surface is equal to the surface velocity because of the nonslip condition between fluid and solid. At some distance from the cylinder, the velocity decreases with r, just like in an

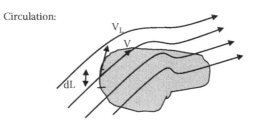

**FIGURE 2.6**
Flow along a closed path.

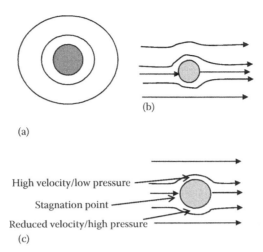

High velocity/low pressure

Stagnation point

Reduced velocity/high pressure

(c)

**FIGURE 2.7**
Velocity field with circulation. (a) Velocity field with circulation around the cylinder, (b) velocity field produced by uniform flow, and (c) superimposed velocity field.

irrotational vortex. Superimposing the velocity field produced for uniform flow (Figure 2.7b) onto a velocity field with circulation around the cylinder (Figure 2.7a), the velocity is reinforced on top and reduced on the bottom (Figure 2.7c). This causes stagnation points to move toward the low-velocity side of the cylinder, resulting in high pressure at the bottom and low pressure on the top and thus creating the lift force. Lift per unit length of an infinitely long cylinder is given by

$$\frac{F_L}{l} = \rho V_0 F$$

where $F_L$ is the lift on the segment $l$, according to ideal-flow theory. For this ideal irrotational flow, there is no drag on the cylinder. For the real-flow case, separation and viscous stresses do produce drag.

### Lift of an Airfoil

Consider the motion of an airfoil through an ideal (nonviscous) fluid. The flow of such a fluid past an airfoil is irrotational, as shown in Figure 2.8(a). The lift and drag forces are zero for an irrotational flow over a cylinder. But for an airfoil, there is a stagnation point on the bottom side near the leading edge, and another on the top side near the trailing edge of the foil. In the real-flow case (viscous fluid), the flow pattern around the front half of the foil as shown in Figure 2.8 is possible. However, the flow pattern in the region of the trailing edge as shown in Figure 2.8a cannot occur. A stagnation point on the upper

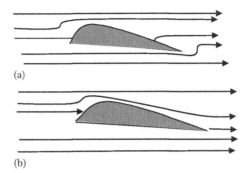

**FIGURE 2.8**
Flow pattern around an airfoil. (a) Flow pattern that is not possible at trailing edge and (b) flow pattern that is possible at trailing edge.

side of the foil indicates that fluid must flow from the lower side around the trailing edge and then toward the stagnation point. Such a flow pattern implies an infinite acceleration of the fluid particles as they turn the corner around the trailing edge of the wing. This is a physical impossibility, and separation will occur at the sharp edge. Therefore, flow from both top and bottom sides of the airfoil in the vicinity of the trailing edge leaves the airfoil smoothly and essentially parallel to these surfaces at the trailing edge, as shown in Figure 2.8b.

To make the theory compatible with the physically observed phenomenon, a circulation with correct intensity around the airfoil must be imposed so that the downstream stagnation point is moved all the way back to the trailing edge of the airfoil, thus allowing the flow to leave the airfoil smoothly at the trailing edge. This is called the Kutta condition, named after the German mathematician and aerodynamicist Martin Wilhelm Kutta. Ideal-flow theory then shows that the magnitude of circulation required to maintain the rear stagnation point at the trailing edge (Kutta condition) of a symmetrical airfoil with a small angle of attack, $\alpha$, is given by

$$\Gamma = \pi c V_0 \alpha$$

where $\Gamma$ is circulation, $c$ is chord length of the airfoil, and $V_0$ is velocity.

Like that of the cylinder, the lift per unit length for an infinitely long wing is

$$\frac{F_L}{1} = \rho V_0 \Gamma$$

The plan form area for the length segment l is

$$F_L = \rho V_0^2 \pi c l \alpha$$

For an airfoil, we define the lift coefficient as

$$C_L = \frac{F_L}{S\rho\left(V_0^2/2\right)}$$

where S is the plan form area of the wing (i.e., the area seen from the top view). With $S = lc$, irrotational flow past a two-dimensional airfoil is

$$C_L = 2\pi\alpha$$

## The Joukowski Transformation for Airfoils

An airfoil is a streamlined structure that will experience as little drag force as possible while generating the maximum lift force when moving against wind. Hence, it must have a smooth surface with no sharp edges facing the wind and a sharp trailing edge so as to allow reattachment of the flow as soon as it leaves the surface. Such a profile can be obtained by using a conformal mapping function called the Joukowski transformation.

A mapping in the plane is said to be angle preserving or conformal if it preserves angles between oriented curves in magnitude and sense. Mapping by an analytical function $f(z)$ is conformal except at $f'(z) = 0$.

Consider the mapping function defined by

$$w = z + \frac{1}{z}$$

In order to verify whether the mapping is conformal, we examine the derivative of w with respect to z, given by

$$w' = 1 - \frac{1}{z^2} = \frac{(z+1)(z-1)}{z^2}$$

The derivative is continuous except at points $z = 1$ and $z = -1$. Hence, the mapping is conformal at all points except at $z = \pm1$. These points correspond to $w = \pm2$, respectively. From the mapping function, we can express z in terms of w as

$$z = \frac{w}{2} \pm \sqrt{\frac{w^2}{4} - 1} = \frac{w}{2} \pm \frac{1}{2}\sqrt{(w+2)(w-2)}$$

Hence, the points w = ±2 are the branch points of the first order of z = z(w). Except at w = ±2, each value of w has two values of z. Consequently, the function w = w(z) maps the z-plane onto a two-sheeted Riemann surface; the sheets are connected from w = −2 to w = 2. Further, this is a one-to-one mapping. Consider circles in the z-plane defined by z = re^{iθ}. Curves having r = constant are circles in the z-plane, and θ = constant are radial lines. Let us examine what these curves look like in the w-plane (Figure 2.9):

$$w = u + iv = re^{i\theta} + \frac{1}{r}e^{-i\theta} = \left(r + \frac{1}{r}\right)\cos\theta + i\left(r - \frac{1}{r}\right)\sin\theta$$

The real and imaginary parts of w are

$$u = \left(r + \frac{1}{r}\right)\cos\theta$$

$$v = \left(r - \frac{1}{r}\right)\sin\theta$$

Eliminating θ between the two equations, we can write

$$\frac{u^2}{a^2} + \frac{v^2}{b^2} = 1$$

where

$$a = r + \frac{1}{r}$$

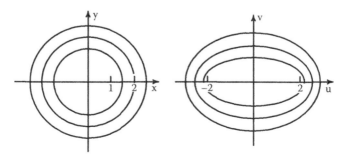

**FIGURE 2.9**
The Joukowski transformation.

and

$$b = \left| r - \frac{1}{r} \right|$$

The circles $r$ = constant in the z-plane are thus mapped onto ellipses in the w-plane with the principal axes of lengths of $2a$ and $2b$ along the u- and v-axes, respectively. Since $a^2 - b^2 = 4$, independent of $r$, these ellipses are confocal with foci at $w = \pm 2$. The unit circle $r = 1$ in the z-plane maps into the line segment from $w = -2$ to $w = 2$ in the w-plane. Except for $r = 1$, circles with radius $r$ and $1/r$ map onto the same ellipse in the w-plane, corresponding to the two sheets of the Riemann surface. Hence, the interior of the unit circle, $|z| = 1$, corresponds to one sheet, and the exterior to the other.

Further, eliminating $r$ between u and v, we get the relation

$$\frac{u^2}{\cos^2 \theta} - \frac{v^2}{\sin^2 \theta} = -4$$

The lines $\theta$ = constant are thus mapped onto the hyperbolas, which are the orthogonal trajectories of these ellipses. The real axis, corresponding to $\theta = 0$ and $\theta = \pi$, are mapped onto the part of the real axis from $w = 2$ to $\infty$ and then from $w = -\infty$ to $-2$. The y-axis corresponding to $\theta = \pi/2$ is mapped onto the v-axis. Any other pair of lines, $\theta = \theta_0$ and $\theta = \theta_0 + \pi$, are mapped onto the two branches of the same hyperbola.

The complete w-plane corresponds to the interior or the exterior of the unit circle, $|z| = 1$.

Since $z = \pm 1$ are branch points in the z-plane, the mapping of curves in the z-plane close to these points is not conformal. Curves near these branch points have a sharp edge with interior angle zero in the w-plane when mapped. If a circle in the z-plane passes through one of these branch points, its mapping onto the w-plane is smooth everywhere except at the point corresponding to the branch point, where it has a sharp edge giving the shape of an airfoil. These are called Joukowski airfoils. In Figure 2.10, a circle, C, is

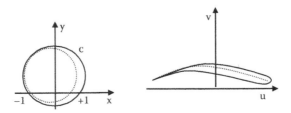

**FIGURE 2.10**
Airfoil shape using the Joukowski transformation.

chosen in the z-plane, which passes through $z = -1$, and $z = 1$ is within the circle. When this is mapped onto the w-plane using the Joukowski transformation, we obtain an airfoil as shown.

### Example 2.1

Transform a circle in the z-plane having a center at $(0.25 + 0.25i)$ and passing through $(-1 + 0i)$ onto the w-plane, using the Joukowski transformation, $w = z + \dfrac{1}{z}$.

### Solution

The radius of the circle is obtained as

$$r = \sqrt{1.25^2 + 0.25^2} = 1.275$$

In the z-plane, the circle is defined as (see Figure 2.11)

$$z = \left(\sqrt{0.25^2 + 0.25^2}\right)e^{i\pi/4} + re^{i\theta} = (0.25 + r\cos\theta) + i(0.25 + r\sin\theta)$$

Consequently, we get

$$w = z + \frac{1}{z}$$

$$= (0.25 + r\cos\theta) + i(0.25 + r\sin\theta) + \frac{1}{(0.25 + r\cos\theta) + i(0.25 + r\sin\theta)}$$

$$= u + iv$$

Plot w corresponding to $r = 1.275$ and $\theta$ varying from 0 to $2\pi$ (see Figure 2.12).

---

## Dimensional Analysis of Force Experienced by a Solid Body in a Flow

The force acting on a body placed in a flow depends on the relative velocity of the fluid with respect to the body. The aerodynamic force has two parts: the lift, which is normal to the direction of the wind, and the drag force, which is due to the skin friction or the shearing effects and is tangential to

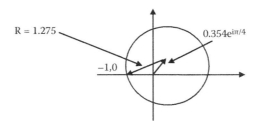

**FIGURE 2.11**
The Joukowski transformation of a circle.

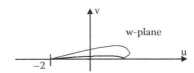

**FIGURE 2.12**
The Joukowski transformation yielding an airfoil shape.

the surface of the body while the other drag forces are in the direction of the wind.

There are four types of drag: skin friction drag, form drag, induced drag, and wave drag. They can also be categorized in two ways: parasitic and induced. The sum of all the components of drag makes up the total drag force. Induced drag or drag due to lift is a small amount of excess (lift) force generated in the opposite direction of the motion. Wave drag generally only occurs when an airplane is flying near the speed of sound (transonic) or faster (supersonic). The form drag, also called pressure drag, is affected by the shape of the body. A smooth, streamlined shape generates less form drag than a blunt body. Automobiles are streamlined to increase gas mileage. Another type of drag, called interference drag, is a component of parasitic drag, which is caused by vortices. Whenever two surfaces meet at a sharp angle, the airflow has a tendency to form a vortex that contributes to drag.

In order to determine the parameters that influence the forces acting on the body when moving against wind, a dimensional analysis can be made.

A real fluid is viscous and compressible. But, if the speed of the flow is much less than the speed of propagation of sound, density variation of the fluid due to the motion of the body is so small that the fluid may be regarded as incompressible. For fluids like water and air, the effects of viscosity are felt only in a thin layer (boundary layer), and outside the layer the fluid may be regarded as nonviscous. A nonviscous and incompressible fluid is called a perfect fluid.

Force exerted by the fluid on the body depends only on the relative velocity between them. Aerodynamic force consists of two components:

1. Pressure force normal to the surface of the body
2. The skin friction or shearing force tangential to the surface of the body

In order to determine the parameters influencing the aerodynamic force on a body, a dimensional analysis can be made. The force, F, is a function of the following parameters:

$$F = f(l, \alpha, \rho, U, \mu, \omega, c)$$

where
l = characteristic length
$\alpha$ = attitude angle
$\rho$ = density of the fluid
$\mu$ = viscosity of the fluid
$\omega$ = frequency of periodic oscillation
$c = \sqrt{\dfrac{p_0 \gamma_0}{\rho}}$ velocity of sound propagation in fluid

Consider the following dimensional table:

| F | l | $\alpha$ | $\rho$ | U | $\mu$ | $\omega$ | c |
|---|---|---|---|---|---|---|---|
| $\dfrac{M\ L}{T^2}$ | L | | $\dfrac{M}{L^3}$ | $\dfrac{L}{T}$ | $\dfrac{M}{L\ T}$ | $\dfrac{1}{T}$ | $\dfrac{L}{T}$ |

The dimensions of the viscosity can be obtained from the expression for the shear stress, since

$$\tau = \mu \frac{du}{dy}, \quad [\tau] = \left[ \frac{M\ L}{T^2} \cdot \frac{1}{L^2} \right] = [\mu] \left[ \frac{L\ 1}{T\ L} \right],$$

$$[\mu] = \left[ \frac{M}{L\ T} \right]$$

We have
n = number of variables = 8
m = number of dimensions = 3
k = n − m = 5, number of nondimensional parameters

We select l, ρ, and U as repeating variables and formulate the nondimensional parameters from the remaining five variables as follows:

1. The first nondimensional parameter corresponds to μ and is given by

$$\pi_1 = 1^a \rho^b U^c \cdot \mu$$

where a, b, and c are powers of the repeating variables and are to be determined as follows: On the left-hand side, $\pi_1$ is a nondimensional quantity and hence

$$\begin{bmatrix} L^0 & M^0 & T^0 \end{bmatrix} = [L]^a \left[\frac{M}{L^3}\right]^b \left[\frac{L}{T}\right]^c \left[\frac{M}{L\ T}\right]$$

Equating the powers of the L, M, and T, we have

$$a - 3b + c - 1 = 0$$

$$b + 1 = 0$$

$$-c - 1 = 0$$

Solving for a, b, and c, we get

$$a = -1, b = -1, \text{ and } c = -1$$

Accordingly, we have

$$\pi_1 = \frac{\mu}{1\rho U}$$

which can be expressed in the form

$$\pi' = \frac{U1\rho_1}{\mu} = R - \text{Reynold's number}$$

2. The second nondimensional parameter corresponds to ω and hence

$$\pi_2 = 1^a \rho^b U^c \cdot \omega$$

Hence,

$$\begin{bmatrix} L^0 & M^0 & T^0 \end{bmatrix} = [L]^a \left[ \frac{M}{L^3} \right]^b \left[ \frac{L}{T} \right]^c \left[ \frac{1}{T} \right]$$

Equating powers of L, M, and T on both sides, we get

$$a - 3b + c = 0$$

$$b = 0$$

$$-c - 1 = 0$$

Solving for a, b, and c, we get

$$a = 1, b = 0, \text{ and } c = -1$$

Hence,

$$\pi_2 = \frac{1\omega}{U} = k, \text{ Strouhal number}$$

3. The third nondimensional parameter corresponds to the speed of sound, c, and

$$\pi_3 = 1^a \rho^b U^d \cdot c$$

Hence, we have

$$\begin{bmatrix} L^0 & M^0 & T^0 \end{bmatrix} = L^a \left[ \frac{M}{L^3} \right]^b \left[ \frac{L}{T} \right]^d \left[ \frac{1}{T} \right]$$

Equating powers of L, M, and T on both sides we get

$$a - 3b + d + 1 = 0$$

$$b = 0$$

$$-d - 1 = 0$$

Solving for a, b, and d, we get

$$a = 0, b = 0, d = -1$$

Hence,

$$\pi_3 = \frac{c}{U} \quad \text{or} \quad \pi_3' = \frac{U}{c}, \text{Mach number}$$

4. The fourth nondimensional parameter corresponds to the force, F, and hence

$$\pi_4 = 1^a \rho^b U^c \cdot F$$

Accordingly, we have

$$\left[ L^0 \; M^0 \; T^0 \right] = [L]^a \left[ \frac{M}{L^3} \right]^b \left[ \frac{L}{T} \right]^c \left[ \frac{M}{T^2} \; \frac{L}{} \right]$$

Equating the powers of L, M, and T on both sides, we get

$$a - 3b + c + 1 = 0$$

$$b + 1 = 0$$

$$-c - 2 = 0$$

Solving for a, b, and c, we get

$$a = -2, b = -1, \text{ and } c = -2$$

Hence,

$$\pi_4 = \frac{F}{1^2 \rho U^2}, \text{nondimensional force}$$

5. $\alpha$ is nondimensional.

Dimensional analysis shows that, for geometrically similar bodies, the force, F, acting on the body can be expressed as

$$F = f\left( \alpha, \; \frac{Ul\rho}{\mu}, \; \frac{\omega l}{U}, \; \frac{U}{c} \right) \frac{1}{2} \rho l^2 U^2$$

where f is a function of the variables inside the parentheses.

The variables are

$R = \dfrac{Ul\rho}{\mu}$, Reynolds number

$k = \dfrac{\omega l}{U}$, Strouhal number (or reduced frequency)

$M = \dfrac{U}{c}$, Mach number

$q = \dfrac{1}{2}\rho U^2$, dynamic pressure

$\upsilon = \dfrac{\mu}{\rho}$, kinematic viscosity

$c = \sqrt{\dfrac{p_o \gamma_c}{\rho}}$, where $p_o$ is the static pressure and $\gamma_c$ is the ratio of specific heats at constant pressure to one at constant volume $\approx 1.4$

For air,

$\rho = 1.21 \text{ kg/m}^3$ at 15°C; at 760 mm of Hg
$P_o = 1.013 \times 10^{-5} \text{ N/m}^2$
$\upsilon = 1.4549 \times 10^{-5} \text{ m}^2/\text{sec.}$

Hence, the velocity of sound in air is

$$c = \sqrt{\frac{1.013 \times 10^5 \times 1.4}{1.21}} = 342.4 \text{ m/sec (in air)}$$

In aeroelasticity, two components of force and a moment are of concern:

1. Lift = L = force perpendicular to the direction of motion
2. Drag = D = force in the direction of motion—positive when the force acts in the downstream direction
3. Pitching moment = M = moment about an axis perpendicular to both the direction of motion and the lift vector—positive when it tends to raise the leading edge of the body

For an airplane wing, the mean chord, c, of the wing is usually taken as the characteristic length and the wing area, S, as the characteristic area.

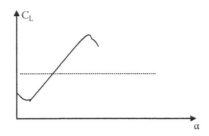

**FIGURE 2.13**
Lift coefficient.

The three primary airplane coefficients are

1. $C_L$ = lift/(qS), lift coefficient
2. $C_D$ = drag/(qS), drag coefficient
3. $C_M$ = (pitching moment/qSc), pitching moment coefficient

$C_L$, $C_D$, and $C_M$ are functions of the Reynolds number (Re), Mach number (M), Strouhal number (k), shape of the body, and its attitude with respect to the flow. For an airfoil, the attitude is described by the angle between the direction of motion, U, and a reference axis called the chord line. This is also referred to as angle of attack ($\alpha$).

In a steady flow of an incompressible fluid, the Strouhal number and the Mach number both vanish and hence $C_L$, $C_D$, and $C_M$ depend on Re and $\alpha$ alone.

Considering incompressible flow, the variation of $C_L$ with $\alpha$ is illustrated in Figure 2.13. When $\alpha$ is small, $C_L$ increases linearly with $\alpha$ and the proportionality constant is called the "lift curve slope," which is denoted by "a." When $\alpha$ becomes large, the lift curve slope "a" is nearly independent of the Reynolds number.

# 3

---

## *Static Aeroelasticity*

---

## Introduction

Aeroelasticity involves the mutual interaction among inertial, elastic, and aerodynamic forces.

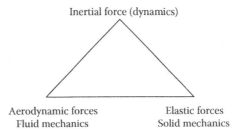

The interaction between aerodynamic forces and elastic forces results in a static aeroelastic phenomenon. But when inertia forces are also present along with the other two, the result is a dynamic aeroelastic phenomenon.

A cross section of an airfoil is shown in Figure 3.1. Points of interest germane to a discussion of static aeroelastic phenomenon are shown in Figure 3.1. In order to explain divergence, which is a static aeroelastic phenomenon, it is required to develop a very simple one degree of freedom model for the airfoil. Such a model will be developed after discussing the different points of interest on the airfoil cross section.

The aerodynamic center is the point on the airfoil about which the aerodynamic moment is independent of angle of attack, $\alpha$. The moment coefficient about this center does not vary with $C_L$. This is located close to the one-quarter chord point for subsonic flow in an incompressible fluid. The aerodynamic center moves to the midchord point when the flow is entirely supersonic.

Strip Theory: Spanwise lift distribution corresponding to a twisted airfoil is of great importance in aeroelasticity. Strip theory assumes that the local lift coefficient, $C_L(y)$, is proportional to the local geometric angle of attack $\alpha(y)$:

$$C_L(y) = a\alpha(y)$$

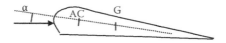

**FIGURE 3.1**
Cross section of an airfoil.

The effect of finite span is then calculated using a value of "a" corrected for aspect ratio.

The shear center is a point through which a shear force produces a pure or "torsion-free" bending.

The elastic axis is the locus of shear centers of the cross section of a beam.

The concepts of shear center and elastic axis lose their simplicity or usefulness for structures other than simple cylindrical beams. The terms flexural center, center of twist, and flexural line are used instead for complex sections such as airfoils.

The flexural center of a cross section is a point at which a shear force can be applied without producing a rotation of that section in its own plane.

The center of twist is a point in a cross section that remains stationary when a torque is applied in that section.

A flexural line is a curve on which loading can be applied so that no twist occurs at any section of the beam.

## Determination of Shear Center in a Thin-Walled Section

An airplane wing is essentially a monocoque structure with a thin-skin frame being reinforced by longitudinal and transverse stiffeners. This construction is in order to reduce the weight of the structure while satisfying the strength requirements. In order to study the torsional deformation of such a thin-walled section as an airplane wing, it is necessary to locate the shear center of the wing section and determine the elastic axis of the wing.

Pure bending couples acting in a principal plane of a beam produce bending deformation in that plane only. Under such pure bending conditions, the beam does not develop any shear stresses. The internal stress resultants on any section form a couple equal and opposite to the externally applied couple at either end of the beam. However, if the beam is subjected to transverse loading, then shearing stresses are also present in addition to the normal stresses. The resultant of these shear stresses on any section is a force parallel to the plane of loading, but it may not act in the plane of loading itself. Hence, the result may be a twisting action on the beam so that cross sections may rotate about the longitudinal axis during bending. If the external loading is applied in the same axial plane as that in which the shear stress resultant acts, then the beam will undergo simple bending without twisting.

In order to determine the plane of loading that does not introduce a twist while causing bending, it is necessary to examine the shear stress distribution over a cross section. Consider a beam that has a symmetric cross section about an axis as shown in Figure 3.2. Let the external load be applied at a distance, e, from the principal plane, xy, which results in simple bending without twisting. The neutral axis of the cross section coincides with the z-axis. The normal stress σ at any point in the cross section will be proportional to the distance, y, from the neutral axis. The shear stresses τ over the flanges in the y direction will follow a parabolic law and the shear stress in the thin horizontal web is negligible.

The two flanges take practically all the shear. Considering the flanges as separate beams with moments of inertia $I_1$ and $I_2$, their curvatures in bending will be same if the load, P, is distributed between them so that

$$\frac{P_1}{P_2} = \frac{b_2}{b_1} = \frac{I_1}{I_2} = \frac{V_1}{V_2}$$

where
  $b_1$ and $b_2$ are the distances to the load from the two flanges
  $V_1$ and $V_2$ are the shear forces in the flanges
  $P = P_1 + P_2$

Thus in order to result in simple bending, the load must be applied between the centroid, C, and the stiffer flange. The point of application of the load, O, is called the shear center of the cross section.

Consider a beam with a thin-walled cross section as shown in Figure 3.3.

The x-axis is along the beam axis, and any cross section is in the xz-plane. The z-axis is the neutral axis of the cross section. The normal stress in any longitudinal fiber of the beam at a distance, y, from the neutral axis is given by

$$\sigma = \frac{M_x y}{I_z}$$

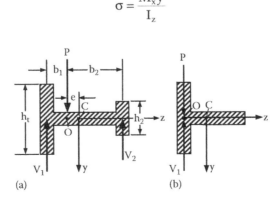

**FIGURE 3.2**
Beam with a thin-walled cross section. (a) Thin-walled beam and (b) beam cross section.

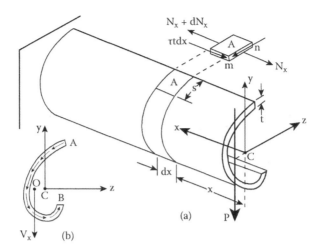

**FIGURE 3.3**
Finding the shear center of a channel section beam. (a) Flanged beam and (b) simplified model.

Forces acting on an element of the beam are shown in Figure 3.3. For the equilibrium of the element in the x direction,

$$\tau t\, dx = dN_x$$

But from the expression for the normal stress $\sigma$, we get

$$N_x = \frac{M_x}{I_z} \int_0^s y\, dA$$

and

$$N_x + dN_x = \frac{M_x + (dM_x/dx)dx}{I_z} \int_0^s y\, dA$$

Hence, we get

$$dN_x = \frac{(dM_x/dx)dx}{I_z} \int_0^s y\, dA$$

Recognizing that $dM_x/dx = V_x$ and that this is the shear force at x, substituting for $dN_x$ in the condition for equilibrium in the x direction, we get an expression for the shear stress at any point in the wall at a distance, s, from the free edge as

$$\tau = \frac{V_x}{I_z t} \int_0^s y \, dA$$

The shear stress is obtained as a function of s, and hence this horizontal shear stress between adjacent fibers varies with the distance s. It is maximum at the neutral plane and zero at each of the free edges.

There must be the same shear stress distribution in the plane of the beam cross section at x, as shown in Figure 3.3. Since the wall thickness of the beam is very small, the shear stresses in the plane of the cross section are tangential. Hence, the shear stress distribution along the centerline AB of the cross section is given by

$$\tau t = \frac{V_x}{I_z} \int_0^s y \, dA$$

This shear stress distribution is called the shear flow.

For the equilibrium of the beam segment between any cross section at x and the free end, the elemental shear force $\tau t ds$ over any cross section must reduce to a force $V_x$ parallel to the y-axis. This shear force will act through a point, O, such that its moment about the centroid C is equal to the sum of the moments of the elemental forces about this point. This condition can be used to determine the shear center, O, for the cross section. The beam will undergo simple bending without twist only if the load, P, at any cross section acts through the shear center of that cross section.

### Example 3.1

A load, V, acts on the thin-walled channel section beam in the vertical plane parallel to the web, as shown in Figure 3.4. Find the distance, e, defining the shear center through which the load should act in order to result in simple bending without twist. Given: h = 9.56 in., b = 2.74 in., t = 0.436 in., I = 78.5 in$^4$.

Since the wall thickness, t, is very small, the shear stresses in the cross section are tangential to the centerline everywhere. Hence, shear stresses are horizontal in the flanges and vertical in the web.

Considering the equilibrium of the cross section, the total shear force in the vertical direction is only in the web, and its magnitude must be equal to the total shear force, V, acting on the entire section. The resultant shear force in each flange can be obtained by knowing the shear stress intensity at any point, s, on the flange. The shear stress intensity is given by

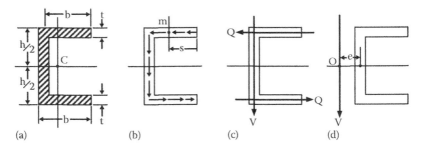

**FIGURE 3.4**
Shear center of a beam of a semicircular cross section. (a) Channel section beam, (b) tangential shear stresses, (c) total shear force on vertical web, and (d) line of action of the total shear force.

$$\tau_s = \frac{V}{It}\int_0^s \frac{h}{2}t\,ds = \frac{Vhs}{2I}$$

Integrating over the entire flange, the resultant shear force, Q, for the flange is given by

$$Q = \int_0^b \tau_s t\,ds = \frac{Vht}{2I}\int_0^b s\,ds = \frac{Vhtb^2}{4I}$$

The shear forces on the top and bottom flanges act along their respective middle lines and constitute a couple of magnitude Qh. The vertical load, V; the shear stress resultant in the web; and the couple Qh formed by the shear stress resultants in the two flanges must keep the cross section in equilibrium. The resultant shear force in the web balances the vertical force V. The horizontal shear force resultants in the top and bottom flanges are equal and opposite in direction, and hence cancel each other. For the moment equilibrium, we can take moment of the forces about the midpoint on the web. For bending with no twist, the moment must be equal to zero. Assuming that, under these conditions, the shear force, V, acts at a distance of e from the midpoint of the web, we get

$$Ve = Qh$$

and hence

$$e = \frac{Qh}{V} = \frac{b^2h^2t}{4I}$$

Substituting the values, we get e = 0.952 in.

**Example 3.2**

A thin-walled beam with a semicircular cross section is loaded in a principal plane xy, as shown in Figure 3.5. Find the location of the shear center of the cross section.

**SOLUTION**

The magnitude of the shear stress at a distance $s = r\phi$ is given by

$$\tau = \frac{V}{I_z t} \int_0^s y\, dA = \frac{V}{I_z t} \int_0^\phi r^2 t \cos\psi\, d\psi = \frac{Vr^2}{I_z} \sin\phi$$

The shear stress is maximum at $\phi = \pi/2$. The sum of the moments of the shear force distribution on the cross section about the center, C, is given by

$$T = \int (\tau tr\, d\phi) r = \frac{Vr^4 t}{I_z} \int_0^\pi \sin\phi\, d\phi = \frac{2Vr^4 t}{I_z}$$

It is seen that the horizontal components of the shear forces over the entire cross section cancel and hence the shear force resultant is equal to the vertical load, V. In order to balance the moments created by the shear forces, the load, V, must act at a point, O, at a distance of e from the center of the cross section such that

$$Ve = T = \frac{2Vr^4 t}{I_z}$$

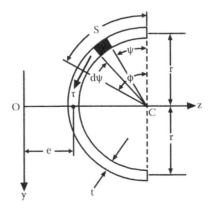

**FIGURE 3.5**
Thin-walled beam with semicircular cross section.

and we get the location of the shear center as

$$e = \frac{2Vr^4t}{VI_z}$$

For the semicircular section, we have the sectional area moment of inertia, $I_z$, given by

$$I_z = \int_0^\pi r^3 t \cos^2 \psi \, d\psi = \frac{\pi r^3 t}{2}$$

After substitution, we get the shear center at e = 1.272 r.

# 4

---

## *Divergence of a Lifting Surface*

---

### Introduction

When a wing is in steady flight, it is subjected to wind loads that cause bending and twisting deformations in the wing. The moment resulting from the lift force on the wing generally tends to twist the wing. As the wing undergoes twisting deformation, the resulting wind pressure distribution causes additional lift force and twisting moment, which in turn increase the twisting deformation. There exists a critical speed at which the elastic stiffness is barely sufficient to hold the wing in a deformed position causing static failure of the wing, and this speed is called divergence speed.

For modern aircraft, divergence speeds are set higher than those of flutter or other aeroelastic instabilities. Hence, divergence speed is of minor importance. However, it is a convenient reference quantity for other aeroelastic phenomena.

In order to understand the behavior of an airplane wing subjected to wind loads during a steady flight, consider a slender flat beam having width much larger than its thickness and a length much larger than the width, as shown in Figure 4.1. Let the force $F_1$ be acting on the front edge of the beam and the force $F_2$ be acting on the rear edge of the beam.

If $F_1 = F_2$, the forces produce a pure moment about the X–X axis causing pure torsion in the beam, without resulting in any bending of the beam axis. Hence, axis X–X is the elastic axis of the beam. However, if $F_2 = 0$, then $F_1$ produces both bending and torsion of the beam. If force $F_1$ acts anywhere between the front edge and the axis X–X, the beam bends up and twists in the counterclockwise direction when looking from the right side of the wing (keeping the leading edge in front) along the axis X–X. This is similar to what happens when the airplane wing is subjected to wind loads. The wind loads are distributed over the wing surface. If we consider a small strip of the wing, the result of the wind loads acts somewhere between the front edge and the elastic axis, generally.

Consider an airplane wing subjected to wind loads during a steady flight. The resultant lift force does not, in general, pass through the elastic axis of

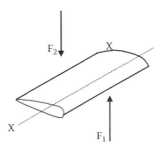

**FIGURE 4.1**
Thin slender flat beam.

the wing, and hence the wing is subjected to both bending and twisting. In reality, an airplane wing is a continuous system with distributed mass and elasticity properties. However, in order to keep the analysis simple and to understand the physics of the divergence phenomenon, it will be convenient to develop a simple model for the airplane wing. As it will be shown later, only the twisting motion affects the divergence phenomenon and the bending of the wing does not have any influence. Hence, it is sufficient to develop a simple single degree of freedom torsional system model for the airplane wing in order to study the phenomenon of divergence.

If a torque is applied at the tip of a cantilever airplane wing, there is a corresponding resisting torque set up in the wing that limits the wing twist, measured at the tip. As the external torque is increased, there is a corresponding increase in the resisting torque as well as in the torsional deformation of the wing. For small deflections, the torsional deformation at the wing tip is proportional to the external torque and hence the wing can be considered as a linear torsional spring. Therefore, a simple torsional model of the wing can be developed by considering a linear torsional spring that resists the external torque due to the lift force on a typical section of the wing. Such a simple model is shown in Figures 4.2 and 4.3.

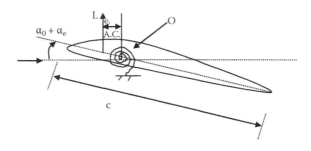

**FIGURE 4.2**
Airfoil.

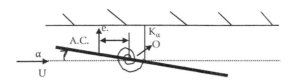

**FIGURE 4.3**
A typical section of airfoil.

Let

$$\alpha = \alpha_o + \alpha_e \tag{4.1}$$

where
$\alpha_o$ = initial angle of attack
$\alpha_e$ = elastic twist of spring

Moment about the elastic axis is

$$M_y = M_{AC} + L \cdot e \tag{4.2}$$

where
$M_y$ = moment about the elastic axis
$M_{AC}$ = moment about the aerodynamic center
L = lift, net vertical force, positive up
e = distance from the elastic axis to aerodynamic center, positive toward nose

From aerodynamic theory,

$$L = C_L \cdot qS \tag{4.3}$$

$$M_{AC} = C_{MAC} \cdot qSc \tag{4.4}$$

where S is the wing area and c is the mean chord length. Lift coefficient varies with $\alpha$ and, using the Taylor series expansion,

$$C_L = C_{LO} \frac{\partial C_L}{\partial \alpha} \cdot \alpha \tag{4.5}$$

The moment coefficient about the aerodynamic center, $C_{MAC}$, does not change with $\alpha$ and hence $C_{MAC} = C_{MACo}$.

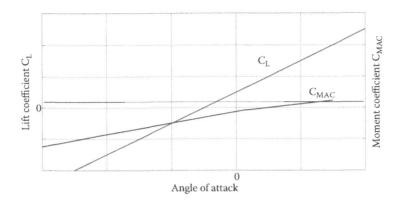

**FIGURE 4.4**
Graph of lift and moment coefficients against angle of attack.

A sketch of lift and moment coefficients plotted against the angle of attack is shown in Figure 4.4. The plots are obtained either from wind tunnel tests or through CFD analysis. For linear spring of constant $K_\alpha$, the elastic moment is $-K_\alpha \cdot \alpha_e$. Hence, for equilibrium,

$$M_y - K_\alpha \cdot \alpha_e = 0$$

Substituting for $M_y$, we get

$$C_{MAC_0} \cdot qSc + \left( C_{L_0} + \frac{\partial C_L}{\partial \alpha} \cdot \alpha \right) \cdot qSe - K_\alpha \cdot \alpha_e = 0 \tag{4.6}$$

Rewriting in the form

$$C_{MAC_0} \cdot qSc + C_{L_0} \cdot qSe + \frac{\partial C_L}{\partial \alpha} \cdot \alpha_0 \cdot qSe = \alpha_e \left( K_\alpha - \frac{\partial C_L}{\partial \alpha} \cdot qS \cdot e \right) \tag{4.7}$$

and solving for $\alpha_e$, we get

$$\alpha_e = \frac{C_{MAC_0} \cdot qSc + C_{L_0} \cdot qSe + \dfrac{\partial C_L}{\partial \alpha} \cdot \alpha_0 \cdot qSe}{K_\alpha - \dfrac{\partial C_L}{\partial \alpha} \cdot qS \cdot e} \tag{4.8}$$

Assuming that $C_{Lo} = 0$ and $C_{MACo} = 0$, we have

$$\alpha_e = \frac{qS}{K_\alpha} \cdot \frac{\dfrac{\partial C_L}{\partial \alpha} \cdot \alpha_o e}{1 - \dfrac{qSe}{K_\alpha} \cdot \dfrac{\partial C_L}{\partial \alpha}} \tag{4.9}$$

Divergence occurs when $\alpha_e \to \infty$ or when

$$1 - \frac{qSe}{K_\alpha} \cdot \frac{\partial C_L}{\partial \alpha} = 0$$

Under these conditions, the divergence dynamic pressure, $q$, is given by

$$q_D = \frac{K_\alpha}{S \cdot e \cdot \dfrac{\partial C_L}{\partial \alpha}} \tag{4.10}$$

Since, $q = 1/2 \, \rho U^2$, the divergence speed is given by

$$U_D = \sqrt{\frac{2q_D}{\rho}} \tag{4.11}$$

We can write a concise form of $\alpha_e$ as

$$\alpha_e = \frac{(q/q_D)\alpha_o}{1 - q/q_D} \tag{4.12}$$

1. If $\alpha_o = 0$, then from Equation 4.7 or 4.8 we have

$$\alpha_e \left( K_\alpha - \frac{\partial C_L}{\partial \alpha} \cdot qSe \right) = 0 \tag{4.13}$$

Hence, either $\alpha_e = 0$, which is trivial case, or

$$\left( K_\alpha - \frac{\partial C_L}{\partial \alpha} \cdot qSe \right) = 0 \tag{4.14}$$

which is the divergence condition. Equation 4.13 is an eigenvalue problem. Divergence requires the consideration of elastic deformation alone.

2. If $\alpha_o \neq 0$ and $\alpha_e \ll \alpha_o$, then from Equation 4.6 with $C_{Lo} = C_{MACo} = 0$, we have

$$eqS \frac{\partial C_L}{\partial \alpha} \cdot \alpha - K_\alpha \cdot \alpha_e = 0 \qquad (4.15)$$

Since $\alpha_e \ll \alpha_o$, we can write

$$eqS \frac{\partial C_L}{\partial \alpha} \cdot \alpha_o - K_\alpha \cdot \alpha_e = 0 \qquad (4.16)$$

Hence,

$$\alpha_e = \frac{eqS \frac{\partial C_L}{\partial \alpha} \cdot \alpha_o}{K_\alpha} \qquad (4.17)$$

and

$$\alpha_e = \frac{q}{q_D} \cdot \alpha_o$$

This is similar to Equation 4.12, when $q \ll q_D$, such that $1 - q/q_D \approx 1$. This means that the dynamic pressure is much smaller than its divergence value. Equation 4.17 cannot predict divergence because it neglects the increase in lift due to elastic twist. However, $q/q_D$ can be usefully thought of as an aeroelastic feedback.

Consider the feedback loop in Figure 4.5.

$$\alpha_e = (\alpha_o + \alpha_e) \cdot q/q_D \qquad (4.18)$$

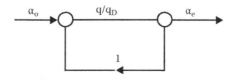

**FIGURE 4.5**
Aeroelastic feedback.

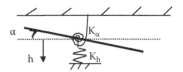

**FIGURE 4.6**
Typical section with both bending and twisting.

From the preceding relation we can get the elastic twist, $\alpha_e$, as

$$\alpha_e = \frac{\dfrac{q}{q_D} \cdot \alpha_o}{1 - \dfrac{q}{q_D}}$$

When the forward loop gain, $(q/q_D)$, exceeds unity, the system is statically unstable. Hence, aeroelasticity can be thought of as the study of aerodynamic and elastic feedback systems. We can also note the similarity of divergence problems to conventional buckling of structures.

Divergence due to both bending and twisting can be seen in Figure 4.6.

Consider the divergence phenomenon in both twisting and bending. The equations of motion are

$$K_h \cdot h + L = 0$$

$$K_\alpha \cdot \alpha_e - Le = 0$$

$$K_h \cdot h + qS \frac{\partial C_L}{\partial \alpha} \cdot \alpha = 0 \tag{4.19}$$

$$K_\alpha \cdot \alpha_e - qSe \frac{\partial C_L}{\partial \alpha} \cdot \alpha = 0 \tag{4.20}$$

$$\alpha = \alpha_o + \alpha_e$$

Hence,

$$K_h \cdot h + qS \frac{\partial C_L}{\partial \alpha} \cdot \alpha_e = -qS \frac{\partial C_L}{\partial \alpha} \cdot \alpha_o$$

$$0 + \left( K_\alpha - qSe \frac{\partial C_L}{\partial \alpha} \right) \alpha_e = qSe \frac{\partial C_L}{\partial \alpha} \cdot \alpha_o$$

$$\alpha_e = \frac{qSe\dfrac{\partial C_L}{\partial \alpha} \cdot \alpha_o}{K_\alpha - qSe\dfrac{\partial C_L}{\partial \alpha}}$$

$$h = \frac{-K_\alpha \cdot qS\dfrac{\partial C_L}{\partial \alpha} \cdot \alpha_o}{K_h \cdot \left(K_\alpha - qSe\dfrac{\partial C_L}{\partial \alpha}\right)}$$

Divergence in twisting and bending occurs at the same speed!

# 5

## Divergence of a Typical Section with a Control Surface

### Introduction

The control surfaces, called ailerons, control the rolling motion of an airplane. When the aileron is displaced downward, the lift over the wing increases and when it is displaced upward, the lift over the wing decreases. In order to maximize the rolling moment, the ailerons on either side of the wing are deflected opposite to each other. But aileron deflection also affects the pitching moment. An aileron deflected downward creates a nose-down aerodynamic pitching moment, which twists the wing in a direction that tends to reduce lift and so the rolling moment. As the elastic stiffness of a wing is independent of flight speed and aerodynamic force varies with $U^2$, there exists a critical speed at which the aileron becomes completely ineffective. This critical speed is called aileron reversal speed. When air speed is above this critical speed, the control is reversed. This means that a downward deflection of an aileron on the starboard wing produces a rolling moment, which moves the starboard wing tip downward instead of upward as intended. Closer to critical speed, aileron control is ineffective.

Consider the addition of a control surface to the typical section as shown in Figure 5.1. Consider also, for simplicity, $\alpha_o = C_{MAC_o} = 0$. The aerodynamic lift is given by

$$L = qSC_L = qS\left(\frac{\partial C_L}{\partial \alpha} \cdot \alpha + \frac{\partial C_L}{\partial \delta} \cdot \delta\right) \tag{5.1}$$

The aerodynamic moment is given by

$$M_{AC} = qScC_{MAC} = qSc\frac{\partial C_{MAC}}{\partial \delta} \cdot \delta \tag{5.2}$$

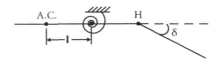

**FIGURE 5.1**
Typical section with control surface.

Note that $\dfrac{\partial C_{MAC}}{\partial \alpha} = 0$, by definition of the aerodynamic center. Both the lift and the aerodynamic moments are positive for nose up. Moment about the hinge line of control surface is

$$H = q S_H c_H C_H = q S_H c_H \left( \frac{\partial C_H}{\partial \alpha} \cdot \alpha + \frac{\partial C_H}{\partial \delta} \cdot \delta \right) \qquad (5.3)$$

where
   $S_H$ is the area of the control surface
   $c_H$ is the chord of the control surface
   $C_H$ is the aerodynamic hinge moment coefficient

$\dfrac{\partial C_L}{\partial \alpha},\ \dfrac{\partial C_L}{\partial \delta},\ \dfrac{\partial C_{MAC}}{\partial \delta},\ \dfrac{\partial C_H}{\partial \alpha},\ \dfrac{\partial C_H}{\partial \delta}$ are the aerodynamic constants, which vary with Mach number and airfoil geometry. Note that $\dfrac{\partial C_H}{\partial \delta}$ is typically negative.

The two equations of static moment equilibrium are

$$eqS \left( \frac{\partial C_L}{\partial \alpha} \cdot \alpha + \frac{\partial C_L}{\partial \delta} \cdot \delta \right) + qSc \frac{\partial C_{MAC}}{\partial \delta} \cdot \delta - K_\alpha \alpha = 0 \qquad (5.4)$$

$$q S_H c_H \left( \frac{\partial C_H}{\partial \alpha} \cdot \alpha + \frac{\partial C_H}{\partial \delta} \cdot \delta \right) - K_\delta (\delta - \delta_0) = 0 \qquad (5.5)$$

where $\delta_0$ is the difference between the angle of zero aerodynamic control deflection and zero twist of the control surface spring. These two equations for $\alpha$ and $\delta$ can be solved using Kramer's rule. Equations 5.4 and 5.5 can be also written as

$$\left( eqS \frac{\partial C_L}{\partial \alpha} - K_\alpha \right) \alpha + \left( eqS \frac{\partial C_L}{\partial \delta} + qSc \frac{\partial C_{MAC}}{\partial \delta} \right) \delta = 0 \qquad (5.6)$$

$$qS_H c_H \frac{\partial C_H}{\partial \alpha} \cdot \alpha + \left( qS_H c_H \frac{\partial C_H}{\partial \delta} - K_\delta \right) \delta = -K_\delta \cdot \delta_0 \qquad (5.7)$$

Divergence occurs when the denominator becomes zero, which is

$$\left( eqS \frac{\partial C_L}{\partial \alpha} - K_\alpha \right)\left( qS_H c_H \frac{\partial C_H}{\partial \delta} - K_\delta \right)$$
$$- \left( eqS \frac{\partial C_L}{\partial \delta} + qSc \frac{\partial C_{MAC}}{\partial \delta} \right) qS_H c_H \frac{\partial C_H}{\partial \alpha} = 0 \qquad (5.8)$$

This is a quadratic equation in q, which has two roots. The lower positive root is the one that corresponds to the divergence speed.

---

## Control Surface Reversal

When $K_\alpha \to \infty$ and $K_\delta \to \infty$, then $\alpha = 0$ and $\delta = \delta_0$.
The lift for the rigid case is

$$L_r = qS \frac{\partial C_L}{\partial \delta} \delta_0 \qquad (5.9)$$

With flexible springs, the lift is

$$L = qS \left( \frac{\partial C_L}{\partial \alpha} \cdot \alpha + \frac{\partial C_L}{\partial \delta} \cdot \delta \right) \qquad (5.10)$$

where $\alpha$ and $\delta$ are determined by solving Equations 5.6 and 5.7. In general, $L < L_r$. The lift may actually become zero or even negative due to aeroelastic effects. This is called control surface reversal.

To simplify matters and to show the essential character of control surface reversal, assume that $K_\delta \to \infty$ and hence $\delta \to \delta_0$ in the equilibrium condition given by Equation 5.7.

From Equation 5.6 we get

$$\alpha = \delta_0 \frac{\dfrac{\partial C_L}{\partial \delta} + \dfrac{c}{e}\dfrac{\partial C_{MAC}}{\partial \delta}}{\dfrac{K_\alpha}{qSe} - \dfrac{\partial C_L}{\partial \alpha}} \qquad (5.11)$$

From Equation 5.10 we have

$$L = qS\left( \frac{\partial C_L}{\partial \delta} \cdot \delta_0 + \frac{\partial C_L}{\partial \alpha} \cdot \alpha \right) = qS\left( \frac{\partial C_L}{\partial \delta} + \frac{\partial C_L}{\partial \alpha} \cdot \frac{\alpha}{\delta_0} \right)\delta_0 \qquad (5.12)$$

Using Equation 5.11 in Equation 5.12 and also using Equation 5.9, we get

$$\frac{L}{L_r} = \frac{1 + q \cdot \dfrac{Se}{K_\alpha} \cdot \dfrac{\partial C_{MAC}}{\partial \delta}\left( \dfrac{\partial C_L}{\partial \alpha} \Big/ \dfrac{\partial C_L}{\partial \delta} \right)}{1 - q \cdot \dfrac{S_e}{K_\alpha} \cdot \dfrac{\partial C_L}{\partial \alpha}} \qquad (5.13)$$

Control surface reversal occurs when $L/L_r = 0$, and hence we have

$$1 + q_R \cdot \frac{Se}{K_\alpha} \cdot \frac{\partial C_{MAC}}{\partial \delta}\left( \frac{\partial C_L}{\partial \alpha} \Big/ \frac{\partial C_L}{\partial \delta} \right) = 0 \qquad (5.14)$$

Consequently,

$$q_R = \frac{-\dfrac{K_\alpha}{Sc}\left( \dfrac{\partial C_L}{\partial \delta} \Big/ \dfrac{\partial C_L}{\partial \alpha} \right)}{\dfrac{\partial C_{MAC}}{\partial \delta}} \qquad (5.15)$$

Typically, $\partial C_{MAC}/\partial \delta$ is negative. Hence, a positive $\delta$ reduces lift. Finally, Equation 5.13 can be written as

$$\frac{L}{L_r} = \frac{1 - q\big/ q_R}{1 - q\big/ q_D} \qquad (5.16)$$

where $q_D$ is given by Equation 4.10 in Chapter 4.

It is interesting to note that even when $K_\delta$ is finite, the reversal dynamic pressure is given by Equation 5.15, with $q_D$ given by the lower root of Equation 5.8. A graphical depiction of Equation 5.16 is given in Figure 5.2.

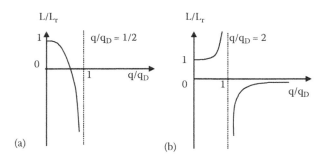

**FIGURE 5.2**
Control surface reversal. (a) $q_D < q_R$ and (b) $q_D > q_R$.

Lift per unit length of airfoil with control surface is

$$L' = qc\left(a\alpha + \frac{\partial C_L}{\partial \delta} \cdot \delta\right)$$

Hence, at critical reversal condition,

$$\frac{dL'}{d\delta} = qc\left(a\frac{\partial \alpha}{\partial \delta} + \frac{\partial C_L}{\partial \delta}\right) = 0$$

$$\frac{\partial \alpha}{\partial \delta} = -\frac{1}{a} \cdot \frac{\partial C_L}{\partial \delta}$$

Check whether $L < L_r$ or $\dfrac{\partial L}{\partial \delta} < \dfrac{\partial L_r}{\partial \delta}$.

# 6

## Dynamic Aeroelasticity

## Introduction

The principal phenomena of interest in dynamic aeroelasticity are

1. Flutter or dynamic instability
2. Response to dynamic loading

In order to understand the dynamic phenomena, including dynamic response, it would be appropriate to become familiar with the vibration of simple systems. Vibration of single degree of freedom and two degrees of freedom systems is discussed in this chapter.

## Vibration Theory

Consider a single degree of freedom vibrating system with a mass, m, suspended by a spring of stiffness, k, as shown in Figure 6.1. The equation of motion is given by

$$m\ddot{x} = -k(x + \delta) + mg$$

But $mg = k \cdot \delta$, where $\delta$ is the static deflection in the spring due to the weight, mg. Hence,

$$m\ddot{x} = -kx - k\delta + mg$$

which results in

$$m\ddot{x} + kx = 0$$

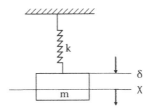

**FIGURE 6.1**
Simple vibrating system.

Denoting $(k/m) = \omega_n^2$, we can write the preceding equation as

$$\ddot{x} + \omega_n^2 x = 0$$

Although this equation is well known to have a harmonic solution, we will derive such a solution and show that the solution is, indeed, harmonic in nature. Assume a power series solution in the form

$$x(t) = \sum_{j=0}^{\infty} a_j t^j$$

Differentiating this twice with respect to time, we get

$$\ddot{x} = \sum_{j=2}^{\infty} a_j \cdot j(j-1) t^{j-2}$$

Substituting for $x$ and $\ddot{x}$ in the equation of motion, we have

$$\sum_{j=2}^{\infty} a_j \cdot j(j-1) t^{j-2} + \omega_n^2 \sum_{j=0}^{\infty} a_j \cdot t^j = 0$$

Expanding the summation, we get

$$(a_2 \cdot 2 \cdot 1 \cdot t^0 + a_3 \cdot 3 \cdot 2 \cdot t + a_4 \cdot 4 \cdot 3 \cdot t^2 + \ldots)$$
$$+ \omega_n^2 (a_0 \cdot t^0 + a_1 \cdot t + a_2 \cdot t^2 + \ldots) = 0$$

This equation is satisfied only when a coefficient of each power of t is equal to zero. Hence, we have

$$\omega_n^2 a_0 + 2a_2 = 0 + a_2 = \frac{-a_0 \omega_n^2}{2}$$

$$\omega_n^2 a_1 + 6a_3 = 0 + a_3 = \frac{-a_1 \omega_n^2}{6}$$

Considering the coefficient of $t^k$, we get

$$\omega_n^2 a_k + (k+2)(k+1)a_{k+2} = 0$$

This provides a recurrence relation among the coefficients as

$$a_{k+2} = \frac{-a_k \omega_n^2}{(k+2)(k+1)}$$

It is necessary to keep the first two coefficients, $a_0$ and $a_1$, as arbitrary and express all other coefficients in terms of $a_0$ and $a_1$ as

$$x(t) = a_0 \left( 1 - \frac{\omega_n^2}{2!} t^2 + \frac{\omega_n^4}{4!} t^4 - \ldots \right)$$

$$+ \frac{a_1}{\omega_n} \left( \omega_n t - \frac{(\omega_n t)^3}{3!} + \frac{(\omega_n t)^5}{5!} - \ldots \right)$$

Quantities inside the parentheses are easily recognized as the Taylor series expansion of $\cos \omega_n t$ and $\sin \omega_n t$, and hence we have

$$x(t) = a_0 \cos \omega_n t + \frac{a_1}{\omega_n} \sin \omega_n t$$

Assuming a solution of the form

$$x(t) = A \cos \omega_n t + B \sin \omega_n t$$

and, expressing the trigonometric terms in their equivalent exponential form, we have

$$x(t) = A \left( \frac{e^{i\omega_n t} + e^{-i\omega_n t}}{2} \right) + B \left( \frac{e^{i\omega_n t} - e^{-i\omega_n t}}{2i} \right)$$

$$X(t) = \frac{A}{2} \left( e^{iw_n t} + e^{-iw_n t} \right) - \frac{B}{2} i \left( e^{iw_n t} - e^{-iw_n t} \right)$$

This can be rewritten in the form

$$x(t) = \frac{1}{2}(A - iB)e^{i\omega_n t} + \frac{1}{2}(A + iB)e^{-i\omega_n t}$$

$$= Ce^{i\omega_n t} + De^{-i\omega_n t}$$

Alternatively, we can assume an exponential solution to the differential equation

$$\ddot{x} + \omega_n^2 x = 0$$

in the form $x = A\,e^{pt}$. Substituting this solution in the differential equation, we get

$$p^2 + \omega_n^2 = 0$$

Hence, we have $p = \pm i\omega_n$, and the solution is obtained as

$$x = A_1 e^{i\omega_n t} + A_2 e^{-i\omega_n t}$$

## Damped Single DOF System

Consider a single degree of freedom (DOF) vibrating system consisting of a mass, spring, and viscous damper, acted upon by a force F(t) as shown in Figure 6.2. The equation of motion is given by

$$m\ddot{x} + c\dot{x} + kx = F(t)$$

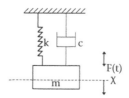

FIGURE 6.2
Damped single degree of freedom system.

Consider the homogeneous case

$$m\ddot{x} + c\dot{x} + kx = 0$$

Assuming an exponential solution, $x = A\,e^{pt}$, and substituting it in the differential equation, we get

$$mp^2 + cp + k = 0$$

Solving this, we get $p$ as

$$p = -\frac{c}{2m} \pm \sqrt{\left(\frac{c}{2m}\right)^2 - \frac{k}{m}}$$

Let $k/m = \omega_n^2$.

## Case 6.1: Critical Damping

When $\left(\dfrac{c}{2m}\right)^2 = \dfrac{k}{m}$, we have $p = -\dfrac{c}{2m}$, a double root, and there are no oscillations.

Hence, $c^2 = 4m^2 \cdot k/m = 4m^2\omega_n^2$. This value of the damping is referred to as critical damping, given by $c = 2\sqrt{km} = 2m\omega_n = c_{cr}$.

Expressing the system-damping coefficient in a nondimensional form by dividing it by the critical damping, we have $c/2m\omega_n = \zeta$. Then $p$ can be written as

$$p = -\zeta\omega_n \pm \sqrt{\zeta^2\omega_n^2 - \omega_n^2}$$

At critical damping, we have $\zeta = 1$ and $p = -\omega_n$, a double root. The corresponding solution is given by

$$x = (A + Bt)e^{-\omega_n t}$$

The behavior is plotted in Figure 6.3. The preceding solution can also be obtained by putting $y_1 = e^{-\omega_n t} = e^{-\frac{c}{2}t}$, $y_2 = u(t) \cdot y_1(t)$.

Substituting into the differential equation and noting that $\dot{y}_1 = -\dfrac{c}{2}y_1$ results in $\ddot{u} = 0$. Solving, we get $u(t) = t$.

**FIGURE 6.3**
Critical damping.

### Case 6.2: Overdamped Oscillations, $\zeta > 1$

When $\zeta > 1$, the two roots of p are given by

$$p = -\zeta\omega_n \pm \sqrt{\zeta^2 - 1}$$

The corresponding solution is given by

$$x = Ae^{\left(-\zeta + \sqrt{\zeta^2 - 1}\right)\omega_n t} + Be^{\left(-\zeta - \sqrt{\zeta^2 - 1}\right)\omega_n t}$$

The resulting oscillations are aperiodic in nature, as shown in Figure 6.4.

### Case 6.3: Damped Oscillations, $0 < \zeta < 1$

In this case, the two roots of p can be written as

$$p = -\zeta\omega_n + i\omega_n\sqrt{1 - \zeta^2}$$

and the corresponding solution is written as

$$x = e^{-\zeta\omega_n t}\left(A\cos\omega_d t + B\sin\omega_d t\right)$$

where $\omega_d = \omega_n\sqrt{1 - \zeta^2}$. This solution corresponds to free oscillations as shown in Figure 6.5.

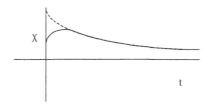

**FIGURE 6.4**
Overdamped system.

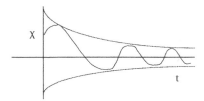

**FIGURE 6.5**
Damped oscillation.

---

## Energy Method

When the system does not dissipate energy during its oscillations and conserves its energy, the system is known as a conservative system. For a conservative system, the total energy remains constant at all times.

The differential equation of such a system can be obtained by equating the variation of energy with respect to time to zero. Let the kinetic energy be denoted by T and the potential energy by U. Then the total energy, $E = T + U$, remains constant at all times. Hence, we have

$$dE/dt = 0$$

and this is the differential equation of motion.

### Example 6.1

Consider the single degree of freedom system shown in Figure 6.6. The kinetic and potential energies of the system are given, respectively, by

$$T = \frac{1}{2}m\dot{x}^2 \text{ and } U = \frac{1}{2}kx^2$$

Hence, the differential equation is obtained as $dE/dt = 0$, which results in

$$m\ddot{x} + kx = 0$$

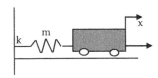

**FIGURE 6.6**
Single degree of freedom system.

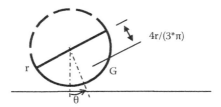

**FIGURE 6.7**
Semicylindrical rod.

### Example 6.2

Obtain the equation of motion of a semicylindrical rod shown in Figure 6.7.
The moment of inertia of the rod about the axis through its center is given by $I_0 = \dfrac{mr^2}{2}$. Transferring it to the centroidal axis, the moment of inertia about the centroidal axis is given by $I_G = \dfrac{mr^2}{2} - m\bar{y}^2$, where $\bar{y} = \dfrac{4r}{3\pi}$.
The x and y coordinates of the centroid when the symmetrical axis of the rod is making an angle $\theta$ with the vertical axis are given by

$$x(\theta) = r\theta - \frac{4r}{3\pi}\sin\theta$$

$$y(\theta) = \frac{4r}{3\pi}(1 - \cos\theta)$$

when the origin of the coordinate system coincides with the equilibrium conditions. The velocities in x and y directions are given by

$$\dot{x} = r\dot{\theta} - \frac{4r}{3\pi}(\cos\theta)\dot{\theta}$$

$$\dot{y} = -\frac{4r}{3\pi}\sin\theta\dot{\theta}$$

For small $\theta$, we can make the following simplifications

$$\dot{x} = r\dot{\theta}\left(1 - \frac{4}{3\pi}\right)$$

$$\dot{y} = -\frac{4r}{3\pi}\theta\dot{\theta}$$

The kinetic energy is given by

$$T = \frac{1}{2} I_G \dot{\theta}^2 + \frac{1}{2} m(\dot{x}^2 + \dot{y}^2)$$

and the potential energy is given by

$$U = mg\left[\frac{4r}{3\pi}(1 - \cos\theta)\right]$$

The differential equation is obtained as $dE/dt = 0$ and is given by

$$\left[mr^2\left(1 - \frac{4}{3\pi}\right)^2 + \frac{mr^2}{2} - mr^2\left(\frac{4}{3\pi}\right)^2\right]\ddot{\theta} + mg\frac{4r}{3\pi}\theta = 0$$

where the nonlinear term has been ignored and $\sin\theta \approx \theta$ for small $\theta$.
Simplifying, we get

$$\left(\frac{3}{2}mr^2 - \frac{8}{3\pi}mr^2\right)\ddot{\theta} + mg\frac{4r}{3\pi}\theta = 0$$

$$\left(\frac{3}{2} - \frac{8}{3\pi}\right)r^2\ddot{\theta} + g\frac{4r}{3\pi}\theta = 0$$

$$\left(\frac{9\pi - 16}{2 \times 3\pi}\right)r\ddot{\theta} + g \cdot \frac{4}{3\pi}\theta = 0$$

$$(9\pi - 16)r\ddot{\theta} + 8g\theta = 0$$

Hence, the natural frequency is obtained as

$$\omega_n^2 = \frac{8g}{(9\pi - 16)r}$$

## Sinusoidal Excitation

Consider a single degree of freedom system acted upon by a harmonic excitation of the form $F(t) = F_0 \cos\omega t$. The equation of motion is given by

$$m\ddot{x} + c\dot{x} + kx = F_0 \cos\omega t$$

This can be written in the form

$$\ddot{x} + 2\zeta\omega_n\dot{x} + \omega_n^2 x = \frac{F_0}{m}\cos\omega t$$

The complete solution is obtained by adding the solution of the homogeneous part of the equation and the particular integral part of the solution.

- *Homogeneous solution:* The solution of the homogeneous part of the equation is obtained by considering the right-hand side of the equation as zero and, consequently,

$$x_c(t) = e^{-\zeta\omega_n t}(A\cos\omega_n t + B\sin\omega_n t)$$

- *Particular integral:* The particular integral part of the solution can be obtained by solving the nonhomogeneous equation. Expressing the excitation in exponential form, we have

$$\cos\omega t = \frac{e^{i\omega t} + e^{-i\omega t}}{2}$$

Consequently, the equation can be expressed in the form

$$\ddot{x} + 2\zeta\omega_n\dot{x} + \omega_n^2 x = \frac{F_0}{2m}(e^{i\omega t} + e^{-i\omega t})$$

The solution of this equation is obtained as

$$x_p = x_{p1} + x_{p2}$$

where

$$x_{p1} = \frac{\left(F_0 / 2m\right)e^{i\omega t}}{\omega_n^2 - \omega^2 + i2\zeta\omega\omega_n}$$

$$x_{p2} = \frac{\left(F_0\big/2m\right)e^{-i\omega t}}{\omega_n^2 - \omega^2 - i2\zeta\omega\omega_n}$$

They can be presented as

$$x_{p1} = \frac{(F_0/2m)e^{+i(\omega t - \phi)}}{D}$$

$$x_{p2} = \frac{(F_0/2m)e^{-i(\omega t - \phi)}}{D}$$

where

$$D = \sqrt{\left(\omega_n^2 - \omega^2\right)^2 + 4\zeta^2\omega_n^2\omega^2}$$

and

$$\phi = \tan^{-1}\frac{2\zeta\omega\omega_n}{\omega_n^2 - \omega^2}$$

The particular integral can also be expressed in terms of trigonometric functions as

$$x_p = \frac{(F_0/m)}{D}\cos(\omega t - \phi)$$

After nondimensionalization of parameters, the particular integral can be expressed as

$$x_p = \frac{\left(F_0/k\right)\cos(\eta\tau - \phi)}{\sqrt{(1 - \eta^2)^2 - 4\zeta^2\eta^2}}$$

where $\eta = \omega/\omega_n$, $\zeta = 2m\omega_n$, BS $\tau = \omega t$.
    The complete solution is given by

$$x = x_c + x_p$$

## Periodic Force

When the system is subjected to a periodic force, the total response of the system can be obtained by expanding the force in a Fourier series to get the response for each term in the series and then superposing these responses together. The differential equation of motion is given by

$$m\ddot{x} + c\dot{x} + kx = F(t)$$

Expressing the force in a Fourier series, we get

$$F(t) = \sum_{n=-\infty}^{\infty} F_n e^{in\omega_0 t}$$

where

$$F_n = \int_{-T/2}^{T/2} F(t) e^{in\omega_0 t} \, dt$$

The particular integral consists of as many terms as those in the Fourier series expansion of $F(t)$.

## Arbitrary Force

When the force acting on the system is arbitrary in nature, as shown in Figure 6.8, it is possible to consider the excitation as a sequence of elementary

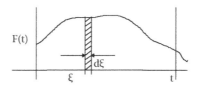

**FIGURE 6.8**
Arbitrary excitation.

impulses of duration $d\tau$. The strength of each impulse at time $\xi$ is equal to $F(\xi)d\xi$. The response of the system at a time, $t$, is given by

$$dx(t) = h(\xi)F(t - \xi)d\xi$$

where $h(t)$ is the unit impulse response of the system. The total response is obtained by superposing the response due to all such elementary impulses. Integrating, we get

$$x(t) = \int_0^t F(t - \xi)h(\xi)d\xi$$

For a damped system subjected to an arbitrary excitation, we have

$$m\ddot{x} + c\dot{x} + kx = F(t)$$

$$h(t) = \frac{1}{m\omega_d}e^{-\zeta\omega_n t}\sin\omega_d t$$

**Example 6.3**

Find the response of an undamped single degree of freedom system to an excitation of the form $F(t) = F_0 \cos \omega t$.

**SOLUTION**

Even though the given $F(t)$ is not arbitrary, this example shows the application of the preceding method.

The equation of motion is given by

$$m\ddot{x} + kx = F_0 \cos \omega t$$

The unit impulse response of the system is given by

$$h(t) = \frac{1}{m\omega_n}\sin\omega_n t$$

Hence, the response is given by

$$x(t) = \frac{F_0}{m\omega_n} \int_0^t \cos\omega(t-\xi)\sin\omega_n\xi \, d\xi$$

$$= \frac{F_0}{2m\omega_n} \int_0^t \left[ \sin\left\{(\omega_n-\omega)\xi+\omega t\right\} + \sin\left\{(\omega_n+\omega)\xi-\omega t\right\} \right] d\xi$$

$$= \frac{F_0}{2m\omega_n} \left[ \frac{-\cos\left\{(\omega_n-\omega)\xi+\omega t\right\}}{\omega_n-\omega} - \frac{\cos\left\{(\omega_n+\omega)\xi-\omega t\right\}}{\omega_n+\omega} \right]_0^t$$

$$= \frac{F_0}{2m\omega_n} \left[ \frac{-\cos\omega_n t + \cos\omega t}{\omega_n-\omega} - \frac{(\cos\omega_n t - \cos\omega t)}{\omega_n+\omega} \right]$$

$$= \frac{F_0}{2m\omega_n} \left[ \frac{-2\omega_n\cos\omega_n t + 2\cos\omega t}{\omega_n^2-\omega^2} \right]$$

$$= \frac{(F_0/k)(\cos\omega t - \cos\omega_n t)}{\left(1 - \omega^2\big/\omega_n^2\right)}$$

Unit impulse response is h(t).
Fourier transform of f(t) is

$$F(\omega) = \frac{1}{2} \int_{-\infty}^{\infty} f(t)e^{-i\omega t} \, dt$$

If $f(t) = \delta(t)$, the impulse force at $t = 0$, then

$$F(\omega) = \frac{1}{2} \int_{-\infty}^{\infty} \delta(t)e^{-i\omega t} \, dt = \frac{1}{2}$$

Let $m\ddot{x} + kx = \delta(t)$. If

$$x = \frac{1}{\pi} \int_{-\infty}^{\infty} x(t)e^{i\omega t} \, d\omega \quad f = \frac{1}{\pi} \int F(\omega)e^{i\omega t} \, d\omega$$

then

$$-m\omega^2 \int x(\omega)e^{i\omega t} \, d\omega + k \int x(\omega)e^{i\omega t} \, d\omega = \int F(\omega)e^{i\omega t} \, d\omega$$

$$\int \left[ (-m\omega^2 + k)x(\omega)e^{i\omega t} - F(\omega)e^{i\omega t} \right] e^{i\omega t} \, d\omega = 0$$

$$x(t) = \int h(\tau) \, f(t - \tau) \, d\tau$$

$$= \int h(\tau) \cdot \frac{1}{\pi} F$$

Hence, $h(t) = \dfrac{1}{\pi} \displaystyle\int_{-\infty}^{\infty} H(\omega)e^{i\omega t} \, d\omega.$

*Fourier transform.* Alternatively,

$$x(t) = \int h(t - \tau)f(\tau)d\tau$$

$$X(\omega) = \frac{1}{2} \int x(t)e^{-i\omega t} \, dt = \frac{1}{2} \int_{-\infty}^{\infty} e^{-i\omega t} \left( \int_{-\infty}^{\infty} h(t - \tau)\delta(\tau)d\tau \right) dt$$

$$= \frac{1}{2} \int_{-\infty}^{\infty} h(t)e^{-i\omega t} \, dt = H(\omega).$$

Hence, h(t) and H($\omega$) are Fourier transform pairs.

---

## Two DOF System

Consider a two DOF system as shown in Figure 6.9. The two degrees of freedom are the vertical translation up and down of the mass, m, and the rotation. The centroidal moment of inertia is J. The beam is of length l and is supported on two springs, $k_1$ and $k_2$. The centroid is located at a distance of $l_1$ from the left end. It is excited by a force, F(t) = F sin$\omega$t, exactly at the centroid. The equations of motion can be written in the matrix form as

$$\begin{bmatrix} m & o \\ o & J \end{bmatrix} \begin{Bmatrix} \ddot{x} \\ \ddot{\theta} \end{Bmatrix} + \begin{bmatrix} (k_1 + k_2) & (k_2 l_2 - k_1 l_1) \\ (k_2 l_2 - k_1 l_1) & (k_1 l_1^2 + k_2 l_2^2) \end{bmatrix} \begin{Bmatrix} x \\ \theta \end{Bmatrix} = \begin{Bmatrix} F(t) \\ 0 \end{Bmatrix}$$

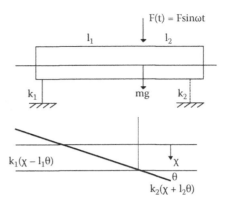

**FIGURE 6.9**
Two degrees of freedom system.

Assuming a solution of the type

$$\left\{\begin{matrix} x \\ \theta \end{matrix}\right\} = \left\{\begin{matrix} X \\ \Theta \end{matrix}\right\} \sin \omega t$$

and substituting in the equation of motion, we get

$$\begin{bmatrix} (k_1 + k_2 - m\omega^2) & (k_2 l_2 - k_1 l_1) \\ (k_2 l_2 - k_1 l_1) & (k_1 l_1^2 + k_2 l_2^2 - J\omega^2) \end{bmatrix} \left\{\begin{matrix} X \\ \Theta \end{matrix}\right\} = \left\{\begin{matrix} F \\ 0 \end{matrix}\right\}$$

The frequency equation is obtained by considering the homogeneous form of this equation, and equating the determinant of the coefficients to zero for nontrivial solution. Consequently, we get

$$\begin{vmatrix} (k_1 + k_2 - m\omega^2) & (k_2 l_2 - k_1 l_1) \\ (k_2 l_2 - k_1 l_1) & (k_1 l_1^2 + k_2 l_2^2 - J\omega^2) \end{vmatrix} = 0$$

Solving the nonhomogeneous equation to obtain the response to the harmonic excitation, we get

$$X = \frac{F(k_1 l_1^2 + k_2 l_2^2 - J\omega^2)}{D}$$

and

$$\Theta = \frac{F(k_1 l_1 - k_2 l_2)}{D}$$

where

$$D = (k_1 + k_2 - m\omega^2)(k_1 l_1^2 + k_2 l_2^2 - J\omega^2) - (k_2 l_2 - k_1 l_1)^2$$

---

## Equations of Motion of a Two DOF Model of an Aircraft Wing

Equations of motion of a two degrees of freedom aircraft wing model are developed here in order to study the flutter phenomenon and also to obtain the response of wings to dynamic loading.

If the deflection is taken to be a consequence of the lift force causing a chordwise rigid rotation about the x-axis and bending of a spanwise elastic axis located at x = 0, then

$$w = h(y) + x\,\alpha(y) \tag{6.1}$$

Virtual work of external forces is (Figure 6.10)

$$\delta W = \iint p \cdot \delta w \cdot dx \cdot dy \tag{6.2}$$

where

$$\delta w = \delta h + x\,\delta \alpha$$

Hence,

$$\delta W = \int \left[ -\int p\,dx \right] \delta h \cdot dy + \int \left[ -\int px\,dx \right] \delta \alpha \cdot dy \tag{6.3}$$

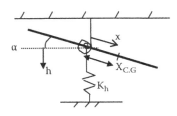

**FIGURE 6.10**
Aircraft wing with two degrees of freedom.

where $L = \int p\,dx$, net vertical force/span.

$M_y = -\int px\,dx$ net moment about the y-axis/span. Then the virtual work is given by

$$\delta W = \int -L\,\delta h\,dy + \int M_y\,\delta\alpha\,dy \qquad (6.4)$$

Lagrange's equations provide the equations of motion as

$$\frac{d}{dt}\frac{\partial L}{\partial \dot{q}_i} - \frac{\partial L}{\partial q_i} = Q_i \qquad (6.5)$$

where

L = T − U is the Lagrangian
t = kinetic energy
U = potential energy
$q_i$ are the generalized coordinates

Let $q_1 = h$ and $q_2 = \alpha$. Let displacement of any point on airfoil be

$$\bar{r} = u\bar{i} + w\bar{k}$$

$$u = x(\cos\alpha - 1) \approx 0$$

since α is small.

$$w = h + x\sin\alpha = h + x\alpha \quad \text{for } \alpha \ll 1$$

Hence, the expressions for the kinetic and potential energies become

$$T = \frac{1}{2}\int(\dot{w}^2 + \dot{u}^2)\rho\,dx \approx \frac{1}{2}\int \dot{w}^2\rho\,dx$$

$$= \frac{1}{2}\int(-\dot{h} - \dot{\alpha}x)^2\,\rho\,dx$$

$$= \frac{1}{2}\dot{h}^2\int\rho\,dx + \frac{1}{2}2\dot{h}\dot{\alpha}\int\rho x\,dx + \frac{1}{2}\dot{\alpha}^2\int\rho x^2\,dx$$

$$= \frac{1}{2}m\dot{h}^2 + \frac{1}{2}2\dot{h}\dot{\alpha}S_\alpha + \frac{1}{2}I_\alpha\dot{\alpha}^2$$

$$U = \frac{1}{2}K_h h^2 + \frac{1}{2}K_\alpha \alpha^2$$

where

$$m = \int \rho\,dx; \quad S_\alpha = \int \rho x\,dx; \quad I_\alpha = \int \rho x^2\,dx$$

$$\delta W = Q_h\,\delta h + Q_\alpha \cdot \delta\alpha$$

Assuming strip theory,

$$Q_h = -L \text{ and } Q_\alpha = M_y$$

per unit span.
   Hence, equations of motion are

$$\frac{d}{dt}(m\dot{h} + S_\alpha\dot{\alpha}) + K_h \cdot h + L = 0 \tag{6.6}$$

$$\frac{d}{dt}(S_\alpha\dot{h} + I_\alpha\dot{\alpha}) + K_\alpha \cdot \alpha - M_y = 0 \tag{6.7}$$

which can be expressed in the form

$$m\ddot{h} + K_h h + S_\alpha\ddot{\alpha} = -L \tag{6.8}$$

$$S_\alpha \cdot \ddot{h} + I_\alpha\ddot{\alpha} + K_\alpha \cdot \alpha = M_y \tag{6.9}$$

*Flutter*

With a steady aerodynamic model, we have

$$L = qS\frac{\partial C_L}{\partial \alpha} \cdot \alpha \text{ and } M_y = eL$$

Hence, equations of motion are

$$m\ddot{h} + S_\alpha \cdot \ddot{\alpha} + K_h \cdot h + qS\frac{\partial C_L}{\partial \alpha} \cdot \alpha = 0 \tag{6.10}$$

$$I_\alpha \cdot \ddot{\alpha} + S_\alpha \cdot \ddot{h} + K_\alpha \cdot \alpha - qSe\frac{\partial C_L}{\partial \alpha} \cdot \alpha = 0 \tag{6.11}$$

Assume a solution of the form

$$h = \bar{h}e^{pt} \tag{6.12}$$

$$\alpha = \bar{\alpha}e^{pt} \tag{6.13}$$

where $p = p_R + i\omega$ and, substituting in Equations 6.10 and 6.11, we obtain

$$\begin{bmatrix} (mp^2 + K_h) & S_\alpha p^2 + qS\dfrac{\partial C_L}{2\alpha} \\ \\ S_\alpha \cdot p^2 & I_\alpha p^2 + K_\alpha - qSe\dfrac{\partial C_L}{\partial \alpha} \end{bmatrix} \begin{Bmatrix} \bar{h} \\ \bar{\alpha} \end{Bmatrix} e^{pt} = \begin{Bmatrix} 0 \\ 0 \end{Bmatrix} \tag{6.14}$$

For nontrivial solutions, the determinant of the matrix on the left-hand side is equated to zero and we have

$$Ap^4 + Bp^2 + C = 0 \tag{6.15}$$

where

$$A = mI_\alpha - S_\alpha^2$$

$$B = m\left(K_\alpha - qSe\dfrac{\partial C_L}{\partial \alpha}\right) + K_h I_\alpha - S_\alpha qS\dfrac{\partial C_L}{\partial \alpha}$$

$$C = K_h\left(K_\alpha - qSe\dfrac{\partial C_L}{\partial \alpha}\right)$$

Hence,

$$p^2 = \dfrac{-B \pm \sqrt{B^2 - 4AC}}{2A} \tag{6.16}$$

With respect to the centroidal axes, we have

$$A = m\,I_G - 0 > 0$$

About noncentroidal axes we have

$$A = m(I_G + d^2 m) - md^2 = m \cdot \int \rho x^2 \, dx - \left(\int \rho x \, dx\right)^2 > 0$$

C is positive if $q < q_D$; that is, if velocity is below divergence speed. B may be either positive or negative. Expanding B,

$$B = mK_\alpha + K_h \cdot I_\alpha - (me + S_\alpha)qS\frac{\partial C_L}{\partial \alpha} \tag{6.17}$$

Hence,

$$B > 0 \text{ if } (me + S_\alpha) < 0 \text{ for all } q.$$

Let us consider the two cases of (1) B > 0 and (2) B < 0.

1. B > 0

We have $p^2$ real and negative if

$$B^2 - 4AC > 0$$

Then p are purely imaginary, representing naturally stable oscillations. On the other hand, if

$$B^2 - 4AC < 0$$

then $p^2$ are complex and at least one value of p will have a positive real part indicating instability. Thus,

$$B^2 - 4AC = 0 \tag{6.18}$$

gives the boundary between neutral stability and instability. The corresponding q is called the flutter value of q. Equation 6.18 is expressed in the form

$$\left[ mK_\alpha + K_h \cdot I_\alpha - (me + S_\alpha)qS\frac{\partial C_L}{\partial \alpha} \right]^2$$
$$- 4\left(mI_\alpha - S_\alpha^2\right) \cdot K_h \left( K_\alpha - qSe\frac{\partial C_L}{\partial \alpha} \right) = 0 \tag{6.19}$$

which can be put in the form

$$D\, q_F^2 + E\, q_F + F = 0 \tag{6.20}$$

where

$$D = \left[ (me + S_\alpha)S \cdot \frac{\partial C_L}{\partial \alpha} \right]^2$$

$$E = \left\{ -2[me + S_\alpha]\left[ mK_\alpha + K_h I_\alpha \right] + 4\left[ mI_\alpha - S_\alpha^2 \right]eK_h \right\}S \cdot \frac{\partial C_L}{\partial \alpha}$$

$$F = [mK_\alpha + K_h I_\alpha]^2 - 4\left[ mI_\alpha - S_\alpha^2 \right] \cdot K_h \cdot K_\alpha$$

If at least one value of $q_F$ is positive, then flutter occurs. If both are positive, the lower value corresponds to flutter. If both are negative, then there is no flutter.

Pines (1958) formulated certain approximate rules for predicting typical section flutter. Accordingly, when $S_\alpha \le 0$, there is no flutter, since for this condition the center of gravity is ahead of the elastic axis. If $S_\alpha > 0$, the flutter speed decreases as $S_\alpha$ increases. $Q_F$ can be raised by reducing $S_\alpha$ by the addition of masses.

2. $B < 0$

B is positive for $q = 0$ and will become negative for large values of q. However, $B^2 - 4AC = 0$ will occur before $B = 0$, since $A > 0$ and $C > 0$. Hence, to determine conditions for flutter, we need to consider only the case of $B > 0$. It can be verified easily that $q > q_F$ when $B < 0$. $C > 0$ only when $q < q_D$. Hence, $q_F < q_D$.

## Quasi-Steady Aerodynamic Theory

Airfoil moving down by a velocity of $\dot{h}$ is identical to the case of an upward wind velocity of $\dot{h}$. The effective angle of attack changes because of the component $\dfrac{\dot{h}}{U}$, as shown in Figure 6.11.

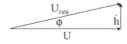

**FIGURE 6.11**
Effective angle of attack.

We have

$$\phi = \tan^{-1}\frac{\dot{h}}{U} \approx \frac{\dot{h}}{U}$$

and hence the corresponding lift force is given by

$$L = qS\frac{\partial C_L}{\partial \alpha}\cdot\left(\alpha + \frac{\dot{h}}{U}\right) \tag{6.21}$$

Then the equation must be written in terms of U as

$$L = \rho\cdot\frac{US}{2}\cdot\frac{\partial C_L}{\partial \alpha}(U\alpha + \dot{h}) \tag{6.22}$$

and the equation will involve terms in p and $p^3$ also, which may be solved numerically.

To show that when $p^2$ is complex then at least one value of p will have a positive real part, consider the following:

$$p^2 = -a \pm ib$$

$$p = c + id$$

$$p^2 = c^2 - d^2 + i2cd = -a \pm ib$$

| | |
|---|---|
| $2cd = \pm b$ | $b^2 = 4c^2d^2$ |
| $-a = c^2 - d^2$ | $d^2 = \dfrac{b^2}{4c^2}$ |
| $-a = c^2 - \dfrac{b^2}{4c^2} = \dfrac{4c^4 - b^2}{4c^2}$ | |

$$4c^4 - b^2 + 4ac^2 = 0$$

$$4c^4 + 4ac^2 - b^2 = 0$$

$$c^2 = \frac{-4a \pm \sqrt{16a^2 + 16b^2}}{8}$$

We have

$$-4a+\sqrt{16a^2+16b^2} > 0, \quad c = \pm\left[\frac{-4a+\sqrt{16a^2+16b^2}}{8}\right]^{1/2}$$

and one value of c is positive.

## Can Flutter Be Seen If Only Torsional Motion Is Considered?

Consider Equation 6.11 only where h motion is dropped:

$$I_\alpha\ddot{\alpha} + K_\alpha\cdot\alpha - qSe\frac{\partial C_L}{\partial\alpha}\cdot\alpha = 0 \tag{6.23}$$

Let $\alpha = \bar{\alpha}e^{pt}$. Then,

$$I_\alpha\cdot p^2 + K_\alpha - qSe\frac{\partial C_L}{\partial\alpha} = 0$$

Hence,

$$p^2 = -\frac{K_\alpha}{I_\alpha} + \frac{qSe}{I_\alpha}\cdot\frac{\partial C_L}{\partial\alpha} \tag{6.24}$$

or

$$p^2 = \frac{K_\alpha}{I_\alpha}\left[-1+\frac{q}{q_D}\right]$$

where $q_D$ is the pressure at divergence. If speed is less than divergence speed, then $q < q_D$ and hence

$$p^2 < 0 \tag{6.25}$$

Hence, $p = \pm i\omega$ and the motion is given by $\alpha = \bar{\alpha}e^{i\omega t}$, which is sinusoidal. Hence, the presence of flutter (below $q_D$) is not indicated by torsional motion only. Above $q_D$, flutter occurs.

## Dynamics of Airfoil

Consider the equation of motion (Equations 6.8 and 6.9).

$$m\ddot{h} + K_h \cdot h + S_\alpha \cdot \ddot{\alpha} = -L \tag{6.26}$$

$$S_\alpha \cdot \ddot{h} + I_\alpha \cdot \ddot{\alpha} + K_\alpha \cdot \alpha = M_y \tag{6.27}$$

When $q \ll q_D$, an approximate analysis for the response may be obtained by considering that the lift and moment are independent of the response parameters h and $\alpha$. Assuming a sinusoidal variation for lift and moment excitations, let

$$L = \bar{L}e^{i\omega t}; \quad M_y = \bar{M}_y e^{i\omega t} \tag{6.28}$$

The resulting steady-state response also will be sinusoidal and may be represented as

$$h = \bar{h}e^{i\omega t}; \quad \alpha = \bar{\alpha}e^{i\omega t} \tag{6.29}$$

Response will have the same frequency as excitation, although with a phase difference (phase included in the complex nature of $\bar{h}$ and $\bar{\alpha}$). Substituting the preceding into Equations 6.26 and 6.27, we get

$$\begin{bmatrix} m(\omega_h^2 - \omega^2) & -S_\alpha \omega^2 \\ -S_\alpha \omega^2 & I_\alpha\left(\omega_\alpha^2 - \omega^2\right) \end{bmatrix} \begin{Bmatrix} \bar{h} \\ \bar{\alpha} \end{Bmatrix} e^{i\omega t} = \begin{Bmatrix} -\bar{L} \\ \bar{M}_y \end{Bmatrix} e^{i\omega t} \tag{6.30}$$

where $\omega_h^2 = K_h/m$, $\omega_\alpha^2 = K_\alpha/I_\alpha \cdot S_\alpha$ couples the h and $\alpha$ motions.

$$\frac{\bar{h}}{\bar{L}} = \frac{-I_\alpha\left(\omega_\alpha^2 - \omega^2\right) + \dfrac{\bar{M}_y}{\bar{L}} S_\alpha \omega^2}{mI_\alpha\left(\omega_h^2 - \omega^2\right)\left(\omega_\alpha^2 - \omega^2\right) - S_\alpha^2 \omega^4}$$

$$H_{hL} = \frac{-[1 - (\omega/\omega_\alpha)^2] + \left(\dfrac{d}{b}\right)\left(\dfrac{x_\alpha}{r_\alpha^2}\right)\left(\dfrac{\omega^2}{\omega_\alpha^2}\right)}{K_h\left\{\left[1 - \dfrac{\omega^2}{\omega_h^2}\right]\left(1 - \dfrac{\omega^2}{\omega_\alpha^2}\right) - \left(\dfrac{x_\alpha^2}{r_\alpha^2}\right)\left(\dfrac{\omega^2}{\omega_\alpha^2}\right)\left(\dfrac{\omega^2}{\omega_h^2}\right)\right\}} \tag{6.31}$$

where

$$d = \bar{M}_y \Big/ \bar{L}; \quad x_\alpha = \frac{S_\alpha}{mb}, \quad r_\alpha^2 = \frac{I_\alpha}{mb^2}$$

and

$$\frac{\bar{h}}{\bar{L}} = H_{hL}\left(\frac{\omega}{\omega_\alpha}; \frac{\omega_h}{\omega_\alpha}; \frac{d}{b}; x_\alpha; r_\alpha\right)$$

The natural frequencies are the roots of the denominator equated to zero.

$$\frac{\omega_1^2}{\omega_h\omega_\alpha}, \frac{\omega_2^2}{\omega_h\omega_\alpha} = \frac{\left(\dfrac{\omega_h}{\omega_\alpha} + \dfrac{\omega_\alpha}{\omega_h}\right) \pm \left\{\left(\dfrac{\omega_h}{\omega_\alpha} + \dfrac{\omega_\alpha}{\omega_h}\right)^2 - 4\left(1 - \dfrac{x_\alpha^2}{r_\alpha^2}\right)\right\}^{1/2}}{2\left(1 - \dfrac{x_\alpha^2}{r_\alpha^2}\right)} \tag{6.32}$$

A similar equation is obtained for response $\dfrac{\alpha}{\bar{L}}$.

**Periodic Motion**

A periodic force, $L(t)$, may be written as

$$L(t) = \sum_{n=-\infty}^{\infty} L_n e^{in\omega_o t} \tag{6.33}$$

where

$$L_n = \frac{1}{T_o} \int_{-T_o/2}^{T_o/2} L(t)e^{-in\omega_o t}\, dt \tag{6.34}$$

with $\omega_o = 2\pi/T_o$ and $T_o$ is period. Response to each term of Equation 6.33 can be summed to obtain the total response as

$$h(t) = \sum_n H_{hL}\left(\frac{n\omega_o}{\omega_\alpha}\right) L_n e^{in\omega_o t} \tag{6.35}$$

## Arbitrary Motion

When a process cannot be described as a sinusoidal process or a periodic process, we can assume that the process is periodic with an infinite period. Accordingly, let $T_o \to \infty$ and define $\omega = n\omega_o$, $\Delta\omega = \Delta n\omega_o$, and $\omega_o = 2\pi/T_o$. Hence,

$$L^*(\omega) = \frac{L_n}{\Delta\omega} = \frac{L_n T_o}{2\pi}$$

is the force per frequency increment.

Then the summation in Equation 6.33 becomes an integral:

$$L(t) = \int_{-\infty}^{\infty} L^*(\omega)e^{i\omega t}\, d\omega \qquad (6.36)$$

where

$$L^*(\omega) = \frac{1}{2\pi}\int_{-\infty}^{\infty} L(t)e^{-i\omega t}\, dt \qquad (6.37)$$

Hence, the response will be

$$h(t) = \int_{-\infty}^{\infty} H_{hL}\left(\frac{\omega}{\omega_\alpha}\right) \cdot L^*(\omega)e^{i\omega t}\, d\omega \qquad (6.38)$$

An alternative form is given by

$$h(t) = \int_{-\infty}^{\infty} I_{hL}(t-\tau)L(\tau)\, d\tau \qquad (6.39)$$

where

$$I_{hL}(t) = \frac{1}{2\pi}\int_{-\infty}^{\infty} H_{hL}\left(\frac{\omega}{\omega_\alpha}\right) \cdot e^{i\omega t}\, d\omega \qquad (6.40)$$

Equation 6.39 is suitable for treating transient motion.

## Laplace Transform

Let

$$L(\tau) = 0 \qquad \text{for } \tau < 0$$
$$I_{hL}(t - \tau) = 0 \qquad \text{for } t - \tau < 0$$

Define $p = i\omega$ and $\omega = -ip$.

Consider the translational motion independent of rotation. Consequently, $S_\alpha = \chi_\alpha = 0$.

$$m\ddot{h} + K_h \cdot h = -L \tag{6.41}$$

Let the force be

$$L = e^{-at} \qquad t > 0$$
$$= 0 \qquad t < 0$$

Hence, we get

$$H_{hL}(\omega) = \frac{-1}{m\left(\omega_h^2 - \omega^2\right)} \qquad \omega_h^2 = K_h / m \tag{6.42}$$

$$I_{hL}(t) = \frac{1}{2\pi} \int_{-\infty}^{\infty} H_{hL} \, e^{i\omega t} \, d\omega \tag{6.43}$$

Hence,

$$I_{hL} = -\frac{1}{m\omega_h} \sin \omega_h t \qquad t > 0$$

$$= 0 \qquad t < 0$$

Since $h(t) = \displaystyle\int_{-\infty}^{\infty} I_{hL}(t - \tau) \, L(\tau) \, d\tau$, we obtain

$$h(t) = -\frac{1}{m\omega_n} \left\{ \frac{\omega_h e^{-at} - \omega_h \cos \omega_h t + a \sin \omega_h t}{a^2 + \omega_h^2} \right\} \tag{6.44}$$

If we include the effect of structural damping, the equation would become

$$m[\ddot{h} + 2\zeta_h\omega_h\dot{h}] + K_h \cdot h = -L \tag{6.45}$$

and the response would be

$$h = -\frac{1}{m\omega_m}\left\{\frac{\omega_h e^{-at} + [-\omega_h\cos\omega_h t + a\sin\omega_h t]e^{-\zeta_h\omega_h t}}{a^2 + \omega_h^2}\right\} \tag{6.46}$$

The impulse function may be determined directly as follows:

$$m\ddot{h} + K_h \cdot h = -L \tag{6.47}$$

Let impulse function be $L(t) = \delta(t)$. Then,

$$m\ddot{I}_{hL} + K_h \cdot I_{hL} = \delta(t)$$

$$\int_0^\varepsilon (m\ddot{I}_{hL} + K_h \cdot I_{hL})dt = -\int_0^\varepsilon \delta(t)dt \tag{6.48}$$

$$m\dot{I}_{hL}\Big|_0^\varepsilon + K_h\int_0^\varepsilon I_{hL}\,dt = -1 \tag{6.49}$$

In the limit when $\varepsilon \to 0$, we get the initial condition:

$$\dot{I}_{hL}(0^+) = -\frac{1}{m}$$

since

$$\dot{I}_{hL}(0) = 0 \quad \text{and} \quad \int_0^\varepsilon I_{hL}\,dt = 0.$$

Integrating Equation 6.49 once more, we get

$$m\,I_{hL}\Big|_0^\varepsilon + K_h \int_0^\varepsilon \left( \int_0^\varepsilon I_{hL}\ d\tau \right) dt = -\int_0^\varepsilon dt$$

Hence     $I_{hL}(0^+) = 0$     $\varepsilon \to 0^+$

Hence, solving Equation 6.47, we get

$$I_{hL} = -\frac{1}{m\omega_n}\sin\omega_h t \quad \text{for} \quad t > 0 \tag{6.50}$$

## Random Motion

Random motion is neither repeatable nor predictable. The actual time variation is of no interest, but certain averages are of interest.

## Mean Values

$$\begin{aligned}
\bar{h} &= \lim_{T\to\infty} \frac{1}{2T} \int_{-T}^{T} h(t)\,dt \\
&= \lim_{T\to\infty} \frac{1}{2T} \int_{-T}^{T} \left( \int_{-\infty}^{\infty} I_{hL}(t-\tau)L(\tau)\,d\tau \right) dt
\end{aligned} \tag{6.51}$$

Interchanging order of integration,

$$= \int_{-\infty}^{\infty} \left\{ \lim_{T\to\infty} \frac{1}{2T} \int_{-T}^{T} L(t-\tau)\,dt \right\} I_{hL}(\tau)\,d\tau$$

$$= \bar{L} \int_{-\infty}^{\infty} I_{hL}(\tau)\,d\tau$$

$$= \bar{L} \cdot H_{hL}(\omega = 0)$$

$$= -\frac{\bar{L}}{K_h}.$$

## Mean Square Value

The correlation function is defined as

$$\phi_{LL}(\tau) = \lim_{T \to \infty} \frac{1}{2T} \int_{-T}^{T} L(t)L(t+\tau)\,dt \tag{6.52}$$

Mean square value is obtained as

$$\bar{L}^2 = \phi_{LL}(\tau = 0)$$

As $\tau \to \infty$, $\phi_{LL}(\tau) \to 0$ since $L(t)$ and $L(t + \tau)$ are uncorrelated for large $\tau$. The autocorrelation function is

$$\phi_{hh}(\tau) = \lim_{T \to \infty} \frac{1}{2T} \int_{-T}^{T} h(t)h(t+\tau)\,dt \tag{6.53}$$

The cross-correlation function is

$$\phi_{hL}(\tau) = \lim_{T \to \infty} \frac{1}{2T} \int_{-T}^{T} h(t)\,L(t+\tau)\,dt \tag{6.54}$$

The power spectral density functions are

$$\phi_{LL}(\omega) = \frac{1}{\pi} \int_{-\infty}^{\infty} \phi_{LL}(\tau)\,e^{-i\omega\tau}\,d\tau \tag{6.55}$$

$$\phi_{LL}(\tau) = \frac{1}{2} \int_{-\infty}^{\infty} \phi_{LL}(\omega)\, e^{i\omega\tau}\, d\omega$$

$$= \int_{0}^{\infty} \phi_{LL}(\omega) \cos \omega\tau\, d\omega$$

(6.56)

Mean square value is

$$\overline{L}^2 = \phi_{LL}(0) = \int_{0}^{\infty} \phi_{LL}(\omega)\, d\omega$$

(6.57)

*To relate the power spectra of input and output,*

$$\phi_{hh}(\tau) = \lim_{T \to \infty} \frac{1}{2T} \int_{-T}^{T} \left\{ \int_{-\infty}^{\infty} L(\tau_1) I_{hL}(t - \tau_1)\, d\tau_1 \right\}$$

$$\times \left\{ \int_{-\infty}^{\infty} L(\tau_2) I_{hL}(t + \tau - \tau_2)\, d\tau_2 \right\} dt$$

(6.58)

Let $t' = t - \tau_1$; $\tau_1 = t - t'$; $t'' = t + \tau - \tau_2$; $\tau_2 = t + \tau - t''$.
Interchanging the order of integration in Equation 6.58, we have

$$\phi_{hh}(\tau) = \int_{-\infty}^{\infty} \int_{-\infty}^{\infty} I_{hL}(t') I_{hL}(t'') \phi_{LL}(\tau + t' - t'')\, dt'\, dt''$$

(6.59)

$$\overline{h}^2 = \phi_{hh}(0) = \int_{-\infty}^{\infty} \int_{-\infty}^{\infty} I_{hL}(t') I_{hL}(t'') \phi_{LL}(t' - t'')\, dt'\, dt''$$

(6.60)

Taking the Fourier transform of $\phi_{hh}(\tau)$, we get

$$\phi_{hh}(\omega) = \frac{1}{\pi} \int_{-\infty}^{\infty} \phi_{hh}(\tau) e^{-i\omega\tau} \, d\tau$$

$$= \frac{1}{\pi} \iiint I_{hL}(t') I_{hL}(t'') \phi_{LL}(\tau + t' - t'') \, e^{-i\omega\tau} \, dt' dt'' d\tau$$

$$= \frac{1}{\pi} \iiint I_{hL}(t') e^{i\omega t'} I_{hL}(t'') e^{-i\omega t''}$$

$$\times \phi_{LL}(\tau + t' - t'') e^{-i\omega(\tau + t' - t'')} \, dt' dt'' d\tau$$

(6.61)

Defining $\tau' = \tau + t' - t''$ and $d\tau' = d\tau$, we get

$$\phi_{hh}(\omega) = H_{hL}(\omega) \, H_{hL}(-\omega) \, \phi_{LL}(\omega) \tag{6.62}$$

Also,

$$\phi_{hL}(\omega) = H_{hL}(\omega) \, \phi_{LL}(\omega)$$
$$\phi_{hh}(\omega) = H_{hL}(-\omega) \, \phi_{hL}(\omega) \tag{6.63}$$

$\phi_{LL}$ is obtained by measurement. Then compute $\phi_{hh}$ and $\overline{h}^2$ where

$$\overline{h}^2 = \int_0^{\infty} \phi_{hh}(\omega) \, d\omega = \int_0^{\infty} |H_{hL}(\omega)|^2 \, \phi_{LL}(\omega) \, d\omega \tag{6.64}$$

**Example 6.4: Airfoil Response to Gust**

Consider translational motion in the vertical direction only. The equation of motion is given by

$$m\ddot{h} + K_h h = -L \tag{6.65}$$

Assuming quasi-steady aerodynamics,

$$L = qS \frac{\partial C_L}{\partial \alpha} \left( \frac{\dot{h}}{U} + \frac{\omega_G}{U} \right) \tag{6.66}$$

where $\left( \dfrac{\dot{h}}{U} + \dfrac{\omega_G}{U} \right)$ is an effective angle of attack, $\alpha$, where $w_G$ is a vertical gust velocity that varies randomly with time.
  Various transfer functions can be defined:

$$\frac{\bar{h}}{L} = H_{hL} = \frac{-1}{m(-\omega^2 + \omega_h^2)}; \quad \omega_h^2 = \frac{K_h}{m} \tag{6.67}$$

is a structural transfer function. Further,

$$\frac{\bar{L}}{h} = H_{Lh} = qS \frac{\partial C_L}{\partial \alpha} \cdot \frac{i\omega}{U} \tag{6.68}$$

is an aerodynamic transfer function (lift due to motion),

$$\frac{\bar{L}}{w_G} = H_{Lw_G} = qS \frac{\partial C_L}{\partial \alpha} \cdot \frac{1}{U}$$

is an aerodynamic transfer function (lift due to gust), and

$$H_{h\omega_G} = \frac{\bar{h}}{\bar{\omega}_G} = \frac{-H_{Lw_G}}{\left[ -\dfrac{1}{H_{hL}} + H_{Lh} \right]} \tag{6.69}$$

is an aerodynamic transfer function (motion due to gust).
  Equation 6.69 is obtained as follows:

$$dL = \frac{\partial L}{\partial h} \cdot dh + \frac{\partial L}{\partial w_G} \cdot dw_G$$

$$dh = \frac{\partial h}{\partial L} \cdot dL$$

$$= \frac{\partial h}{\partial L} \left( \frac{\partial L}{\partial h} \cdot dh + \frac{\partial L}{\partial w_G} \cdot dw_G \right)$$

$$= H_{hL} \left( H_{Lh} \cdot dh + H_{Lw_G} \cdot dw_G \right)$$

$$dh \left( 1 - H_{hL} \cdot H_{Lh} \right) = H_{hL} \cdot H_{Lw_G} \cdot dw_G$$

$$\frac{dh}{d\omega_G} = H_{hw_G} = \frac{H_{Lw_G}}{\dfrac{1}{H_{hL}} - H_{Lh}} \tag{6.70}$$

Using random force–response relations,

$$\overline{h}^2 = \int_{-\infty}^{\infty} \left| H_{hw_G} \right|^2 \phi_{w_G w_G} \, d\omega$$

$$= \int_0^{\infty} \frac{\left( qS \dfrac{\partial C_L}{\partial \alpha} \cdot \dfrac{1}{U} \right)^2}{\left( -m\omega^2 + K_h^2 \right)^2 + \left( qS \dfrac{\partial C_L}{\partial \alpha} \cdot \dfrac{\omega}{U} \right)^2} \cdot \phi_{w_G w_G} \, d\omega \tag{6.71}$$

Define an effective damping constant as

$$\zeta = \frac{qS \dfrac{\partial C_L}{\partial \alpha} \cdot \dfrac{1}{U}}{2\sqrt{mK_h}} \tag{6.72}$$

Thus,

$$\overline{h}^2 = \frac{\left( qS \dfrac{\partial C_L}{\partial \alpha} \cdot \dfrac{1}{U} \right)^2}{m^2} \int_0^{\infty} \frac{\phi_{w_G w_G} \, d\omega}{\left( -\omega^2 + \omega_h^2 \right)^2 + 4\zeta^2 \omega_h^2 \omega^2} \tag{6.73}$$

For small $\zeta$, we get,

$$\overline{h}^2 \approx \frac{qS \dfrac{\partial C_L}{\partial \alpha} \cdot \pi}{K_h} \cdot \frac{\phi_{w_G w_G}(\omega_h)}{U} \tag{6.74}$$

For small $\zeta$,

$$\zeta_{w_G w_G}(\omega) \approx \phi_{w_G w_G}(\omega_h)$$

Typically,

$$\phi_{w_G w_G}(\omega) = \overline{w}_G^2 \frac{L_G}{\pi U} \cdot \frac{1 + 3\left( \dfrac{\omega L_G}{U} \right)^2}{\left[ 1 + \left( \dfrac{\omega L_G}{U} \right)^2 \right]^2} \tag{6.75}$$

# 7

# One-Dimensional Aeroelastic
# Model of Airfoils

## Introduction

In order to develop an aeroelastic model of an aircraft wing with continuously distributed mass and elastic properties, it is crucial to analyze its torsional characteristics. This can be better understood by considering the simple torsion of a bar as discussed next.

## Simple Torsion of a Bar

In developing the equations for the torsion of a simple bar, it is assumed that circular sections remain circular after twist as shown in Figure 7.1.

A disk-like element of the shaft will have a rotation of its bottom cross section with reference to its top through an angle $d\phi$. The element is in a state of pure shear and shearing strain.

$$\gamma = \frac{r \, d\phi}{dx} \tag{7.1}$$

For a shaft twisted by a torque at the end, the angle of twist, $\phi$, is proportional to the distance, $x$, of the cross section from the fixed end, and hence $d\phi/dx$ is constant, which is the angle of twist per unit length, $\theta$. Hence,

$$\gamma = r\theta$$

$$G = \frac{s_s}{\gamma}$$

and hence $s_s = G\gamma = Gr\theta$.

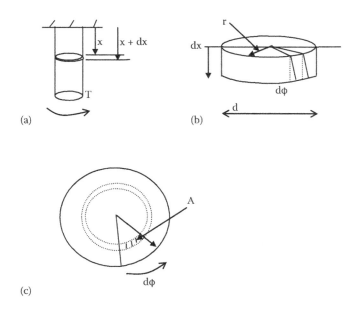

**FIGURE 7.1**
Torsion of a bar of a circular cross section. (a) Bar acted upon by a torque at one end, (b) shear deformation of an element, and (c) element in an annular ring.

The torque over the cross-sectional area is given by

$$T = \int_{\rho_0}^{\rho=d/2} G\theta \cdot r^2 \, dA = G\theta I_p$$

and hence

$$\frac{T}{GI_p} = \theta$$

$$\frac{Tl}{GI_p} = \theta l = \phi$$

$$\frac{T}{GI_p} = \frac{d\phi}{dx}$$

(7.2)

## General Approximations for Aerodynamic Theory

### Strip Theory

In strip theory, the wing span is considered to be infinite with uniform span-wise properties. Lift at any point is assumed to depend only on the down-wash (vertical velocity; normal velocity) at that point and independent of downwash at other points.

### Quasi-Steady Approximations

Aerodynamic forces at any time depend only on the motion of the airfoil at that same time. The lift coefficient is given by

$$C_L = \frac{dC_L}{d\alpha}\left[\alpha + \frac{\dot{h}}{U} + \frac{1}{U}\left(\frac{3}{4}c - x_0\right)\dot{\alpha}\right] \tag{7.3}$$

where $(\dot{h}/U)$ accounts for the effective angle of inclination of the wing due to the vertical translational velocity of the wing, and $x_0$ is the distance from nose to aerodynamic center. The moment coefficient is given by

$$C_{M_{Le}} = -\frac{c\pi}{8U}\cdot\dot{\alpha} - \frac{1}{4}C_L \tag{7.4}$$

Sometimes $C_L$ is approximated to

$$C_L = \frac{dC_L}{d\alpha}\left(\alpha + \frac{\dot{h}}{U}\right) \tag{7.5}$$

### Beam-Rod Model

A flat plate with rigid chordwise sections and $l/c \gg 1$, where $l$ is the span, can be modeled as a beam rod for analysis. The equation of static moment equilibrium is

$$\frac{T}{GJ} = \frac{d\phi}{dy}$$

See Figure 7.2.

Consider the equilibrium of the element. With

$$T = GJ\frac{d\phi}{dy}$$

for equilibrium, we have

$$T + \frac{dT}{dy}dy - T + M_y = \frac{dT}{dy}\cdot dy + M_y = \frac{d}{dy}\left(GJ\cdot\frac{d\phi}{dy}\right)dy + M_y = 0$$

which reduces to

$$\frac{d}{dy}\left(GJ\frac{d\alpha_e}{dy}\right) + M_y = 0 \tag{7.6}$$

where

$$M_y = \frac{dT}{dy} = \text{aerodynamic moment per unit length}$$

$\alpha_e(y)$ = nose-up twist at y
G   = shear modulus
J   = polar moment of inertia

Boundary conditions are given by

$$\alpha_e(0) = 0; \quad GJ\alpha_e'(l) = 0$$

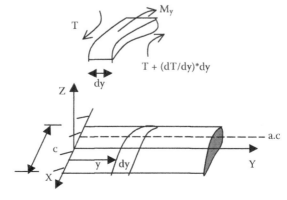

**FIGURE 7.2**
Beam-rod model of an aircraft wing.

Assuming strip theory,

$$M_y = M_{AC} + \text{Le moment/length}$$

$$L = q.c.C_L \text{ lift/length}$$

where

$$C_L(y) = \frac{dC_L}{d\alpha}[\alpha_0(y) + \alpha_e(y)]$$

$$M_{AC} = qc^2 C_{MAC}$$

(7.7)

The equation is

$$\frac{d}{dy}(GJ\alpha_e') + qc^2 C_{MAC} + qce\frac{dC_L}{d\alpha}(\alpha_0 + \alpha_e) = 0$$

$$\alpha_e'' + \frac{qc^2}{GJ}C_{MAC} + \frac{qce}{GJ}\frac{dC_L}{d\alpha}(\alpha_0 + \alpha_e) = 0$$

$$\alpha_e'' + \frac{qce}{GJ}\frac{dC_L}{d\alpha}\cdot\alpha_e = -\frac{qc}{GJ}\left(e\frac{dC_L}{d\alpha}\cdot\alpha_0 + C_{MAC}\cdot c\right)$$

If $\bar{y} = y/l$; then

$$\frac{1}{l^2}\frac{d^2\alpha_e}{d\bar{y}^2} + \frac{qce}{GJ}\frac{dC_L}{d\alpha}\cdot\alpha_e = -\frac{qc}{GJ}\left(e\frac{dC_L}{d\alpha}\cdot\alpha_0 + C_{MAC}\cdot c\right)$$

This equation can be written as

$$\frac{d^2\alpha_e}{d\bar{y}^2} + \lambda^2\alpha_e = k$$

(7.8)

where

$$\lambda^2 = \frac{ql^2}{GJ}\cdot c\frac{\partial C_L}{\partial \alpha}\cdot e$$

(7.9)

$$k = -\frac{qcl^2}{GJ} \cdot \left( e\frac{\partial C_L}{\partial \alpha} \cdot \alpha_0 + C_{MAC} \right)$$

Boundary conditions are

$$\alpha_e(0) = 0 \quad \text{at} \quad \bar{y} = 0$$

$$\frac{d\alpha_e(1)}{d\bar{y}} = 0 \quad \text{at} \quad \bar{y} = 1$$

The general solution of Equation 7.8 is given by

$$\alpha_e = A\cos\lambda\bar{y} + B\sin\lambda\bar{y} + \frac{k}{\lambda^2} \tag{7.10}$$

Applying boundary conditions, we get

$$A + \frac{k}{\lambda^2} = 0 \qquad \lambda(-A\sin\lambda + B\cos\lambda) = 0$$

$$\text{i.e., } A = -\frac{k}{\lambda^2}; \quad B = -\frac{k}{\lambda^2}\tan\lambda$$

Hence,

$$\alpha_e = \frac{k}{\lambda^2}\left(1 - \cos\lambda\bar{y} - \tan\lambda\sin\lambda\bar{y}\right)$$

Divergence occurs when $\alpha_e \rightarrow \infty$; that is,

$$\tan\lambda \rightarrow \infty \quad \text{or} \quad \lambda_1 = \frac{\pi}{2}, \frac{3\pi}{2}, \frac{5\pi}{2}$$

From Equation 7.9 we get

$$q_D = \left(\frac{\pi}{2}\right)^2 \frac{GJ}{1} \bigg/ \left( lce\frac{\partial C_L}{\partial \alpha} \right) \tag{7.11}$$

From the earlier discussion on divergence of a typical section, we recognize that $S = lc$; that is,

$$q_D = \left(\frac{\pi}{2}\right)^2 \frac{GJ}{l} \cdot \frac{1}{lce \dfrac{\partial C_L}{\partial \alpha}} = \frac{K_\alpha}{Se \dfrac{\partial C_L}{\partial \alpha}}$$

We see that equivalent $K_\alpha = \left(\dfrac{\pi}{2}\right)^2 \dfrac{GJ}{l}$.

---

## Eigenvalue and Eigenfunction Approaches

The homogeneous form of Equation 7.8 is given by

$$\frac{d^2\alpha_e}{d\bar{y}^2} + \lambda^2 \alpha_e = 0$$

with the associated boundary conditions of

$$\alpha_e = 0 \quad y = 0 \qquad \frac{\partial \alpha_e}{\partial \bar{y}} = 0 \quad y = 1$$

The solution is

$$\alpha_e = A \cos \lambda \bar{y} + B \sin \lambda \bar{y}$$

Applying the boundary conditions we get,

$$A = 0 \quad \text{and} \quad B\lambda \cos \lambda = 0$$

Since $B \neq 0$ and we want nonzero $\lambda$,

$$\cos \lambda = 0$$

Hence,

$$\lambda = (2m - 1)\frac{\pi}{2} \quad m = 1, 2, 3, \ldots \infty \leftarrow \text{Eigenvalues}$$

$$\alpha_e = \left[ a_m \sin(2m-1)\frac{\pi}{2} \cdot \bar{y} \right] m = 1, 2, 3 \ldots \infty \leftarrow \text{Eigen functions} \qquad (7.12)$$

Consider a nonhomogeneous case,

$$\frac{d^2\alpha_e}{d\bar{y}^2} + \lambda^2 \alpha_e = k \qquad (7.13)$$

Assume that

$$\alpha_e = \sum a_n \alpha_n(\bar{y}) \qquad (7.14)$$

$$k = \sum A_n \alpha_n(\bar{y}) \qquad (7.15)$$

Since

$$\frac{d^2\alpha_e}{d\bar{y}^2} + \lambda^2 \alpha_e = 0$$

is satisfied by $\lambda_n$ and $\alpha_n$, we have

$$\alpha_n'' = -\lambda_n^2 \alpha_n$$
$$\alpha_m'' = -\lambda_m^2 \alpha_m \qquad (7.16)$$

Multiplying the first equation by $\alpha_m$ and the second by $\alpha_n$ and then integrating from 0 to 1, we get

$$\int_0^l \alpha_m \alpha_n'' \, dy = -\lambda_n^2 \int \alpha_m \alpha_n \, dy$$

$$\int_0^l \alpha_m'' \alpha_n \, dy = -\lambda_m^2 \int \alpha_m \alpha_n \, dy \qquad (7.17)$$

That is,

$$\alpha_m \, \alpha_n' \big|_0^l - \int \alpha_m' \alpha_n' \; dy = -\lambda_n^2 \int \alpha_m \alpha_n \; dy$$

$$\alpha_m' \, \alpha_n \big|_0^l - \int \alpha_m' \alpha_n' \; dy = -\lambda_m^2 \int \alpha_m \alpha_n \; dy \tag{7.18}$$

Since $\alpha_m, \alpha_m', \alpha_n, \alpha_n'$ satisfy the boundary conditions at 0 or 1, by subtracting one equation from the other, we have

$$\left(\lambda_m^2 - \lambda_n^2\right) \int \alpha_m \alpha_n \; dy = 0$$

Hence, when

$$\lambda_m = \lambda_n, \int \alpha_m \alpha_n \; dy = \beta$$

$$\lambda_m \neq \lambda_n, \int \alpha_m \alpha_n \; dy = 0$$

This is the orthogonality condition.
Substituting Equations 7.14 and 7.15 into Equation 7.13,

$$\sum_n \left[ a_n \frac{d^2 \alpha_e}{dy^2} + \lambda^2 a_n \right] = \sum A_n \alpha_n \tag{7.19}$$

From the homogeneous case, we have

$$\frac{d^2 \alpha_e}{dy^2} = -\lambda_n^2 \alpha_e \tag{7.20}$$

Substituting Equation 7.20 into Equation 7.19,

$$\sum a_n \left(-\lambda_n^2 + \lambda^2\right) \alpha_n = \sum A_n \alpha_n \tag{7.21}$$

Multiplying Equation 7.21 by $\alpha_m$ and integrating from 0 to 1, we get

$$\left(\lambda^2 - \lambda_n^2\right) a_n \cdot \int \alpha_n^2 \ d\bar{y} = A_n \cdot \int \alpha_n^2 \ d\bar{y} \tag{7.22}$$

Hence,

$$a_n = \frac{A_n}{\left(\lambda^2 - \lambda_n^2\right)} \tag{7.23}$$

Thus,

$$\alpha_e = \sum a_n \alpha_n = \sum \frac{A_n \cdot \alpha_n(\bar{y})}{\left(\lambda^2 - \lambda_n^2\right)} \tag{7.24}$$

$\alpha_e = \infty$   when   $\lambda \to \lambda_1$.

## Galerkin's Method

The equation of equilibrium for a variable property wing is

$$\frac{d}{dy}\left(GJ \frac{d\alpha_e}{dy}\right) + eqc \frac{\partial C_L}{\partial \alpha} \cdot \alpha_e = -eqc \frac{\partial C_L}{\partial \alpha} \alpha_0 - qc^2 C_{MAC_o} \tag{7.25}$$

Using a reference value of $(GJ)_{ref}$, define

$$\gamma = \frac{GJ}{(GJ)_{ref}}; \ k = -\frac{qcl^2}{(GJ)_{ref}}\left[e \frac{\partial C_L}{\partial \alpha} \cdot \alpha_0 + C_{MAC_0} \cdot c\right]$$

$$\lambda^2 = \frac{ql^2 c_{ref}}{(GJ)_{ref}}\left(\frac{\partial C_L}{\partial \alpha}\right)_{ref} e_{ref}; \ \beta = \frac{c}{c_{ref}} \cdot \frac{e}{e_{ref}} \cdot \frac{\partial C_L/\partial \alpha}{\left(\partial C_L/\partial \alpha\right)_{ref}}$$

Then, with $\bar{y} = y/\ell$, we have

$$\frac{d}{dy}\left(\gamma \frac{d\alpha_e}{\partial \bar{y}}\right) + \lambda^2 \alpha_e \beta = k \tag{7.26}$$

Assume

$$\alpha_e = \sum a_n \alpha_n(\bar{y})$$

$$k = \sum A_n \alpha_n(\bar{y})$$

(7.27)

Substituting Equation 7.27 into Equation 7.26, multiplying by $\alpha_m$, and integrating from 0 to 1 along $\bar{y}$:

$$\sum a_n \left\{ \int_0^1 \frac{d}{d\bar{y}} \left( \gamma \frac{d\alpha_n}{d\bar{y}} \right) \alpha_m \, d\bar{y} + \lambda^2 \int_0^1 \beta \alpha_m \alpha_n \, d\bar{y} \right\} = \sum A_n \int_0^1 \alpha_n \alpha_m \quad d\bar{y} \quad (7.28)$$

$\alpha_n$ are eigenfunctions of a uniform wing—in which case,

$$\alpha_n = \sin(2m-1)\frac{\pi}{2}\bar{y}$$

and

$$\int_0^1 \sin^2(2m-1)\frac{\pi}{2}\bar{y} \, d\bar{y} = \frac{1}{2}$$

Here, $\alpha_n$ happens to be a convenient set of functions satisfying the boundary conditions. Hence, Equation 7.28 gives

$$[C_{mn}]\{a_n\} = \{A_n\}$$

(7.29)

and

$$\{a_n\} = [C_{mn}]^{-1}\{A_n\}.$$

The divergence condition corresponds to the first root of

$$|C_{mn}| = 0.$$

## Swept Wing Divergence

Sweep is defined as the angle between the axis perpendicular to the oncoming stream (y-axis) and the elastic axis ($\bar{y}$-axis). Using bending-torsion deformation of a beam rod, the equations of motion are given by

$$\frac{d^2}{d\bar{y}^2}\left( EI\frac{d^2h}{d\bar{y}^2} \right) = -\bar{L}$$

$$\frac{d}{d\bar{y}}\left( GJ\frac{d\alpha_e}{d\bar{y}} \right) + \bar{M}_y = 0$$

Here, h is bending deflection of the elastic axis and $\alpha_e$ is the elastic twist about the elastic axis. Using strip theory aerodynamics with respect to chords perpendicular to the $\bar{y}$ axis, the lift force and moments are

$$\bar{L} = \bar{C}_L \bar{c}\bar{q}$$

$$\bar{M}_y = \bar{L}\bar{e} + \bar{M}_{AC} = \bar{C}_L \bar{c}\bar{q}e + \bar{C}_{MAC}\ \bar{c}^2\bar{q}$$

where $\bar{q} = \dfrac{1}{2}\rho(U\cos\Lambda)^2 = q\cos^2\Lambda$

Consider the velocity diagram shown in Figure 7.3.

The component along the elastic axis is $U\sin\Lambda$. Accordingly, fluid velocity normal to the wing is $\sin\Lambda\dfrac{dh}{d\bar{y}}$. Hence, the effective angle of attack due to this is

$$U\sin\Lambda\frac{dh}{d\bar{y}}\bigg/ U\cos\Lambda = \frac{dh}{d\bar{y}}\tan\Lambda$$

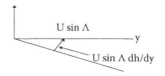

**FIGURE 7.3**
Velocity diagram of a swept wing.

Hence, the total angle of attack is

$$\alpha_T = \alpha_e + \frac{dh}{d\bar{y}} \tan \Lambda$$

and the coefficient of lift is

$$\bar{C}_L = \frac{\partial \bar{C}_L}{\partial \alpha} \cdot \alpha_T(\bar{y})$$

The equations of motion are

$$\frac{d^2}{d\bar{y}^2}\left( EI \frac{d^2 h}{d\bar{y}^2} \right) = -\frac{\partial \bar{C}_L}{\partial \alpha}\left[ \alpha_e + \frac{dh}{d\bar{y}} \tan \Lambda \right] \bar{c}q \cos^2 \Lambda$$

$$\frac{d}{d\bar{y}}\left( GJ \frac{d\alpha_e}{d\bar{y}} \right) + \frac{\partial \bar{C}_L}{\partial \alpha}\left[ \alpha_e + \frac{dh}{d\bar{y}} \tan \Lambda \right] \bar{c}q \cos^2 \Lambda \; \bar{e}$$

$$+ \bar{C}_{MAC} \; \bar{c}^2 q \cos^2 \Lambda = 0$$

## Special Cases

- When $EI \to \infty$, and $h \to 0$, then the torsional equation is similar to that for an unswept wing.
- When $GJ \to \infty$ and $\alpha \to 0$, then the bending equation is

$$\frac{d^2}{d\bar{y}^2}\left( EI \frac{d^2 h}{d\bar{y}^2} \right) + \frac{\partial \bar{C}_L}{\partial \alpha} \sin \Lambda \cos \Lambda \; \bar{c}q \; \frac{dh}{d\bar{y}} = 0$$

Hence, divergence in bending alone is possible if the wing is stiff in torsion. This is not possible for an unswept wing.

To illustrate further, consider a beam with uniform spanwise properties. Then,

$$\frac{d^4 h}{d\bar{y}^4} + \lambda \frac{dh}{d\bar{y}} = 0$$

With $\bar{y} = \bar{y}/l$ and

$$\lambda = \frac{\dfrac{\partial \bar{C}_L}{\partial \alpha} q \bar{c} l^3}{EI} \sin \Lambda \cos \Lambda$$

the boundary conditions are

$$h = \frac{dh}{d\bar{y}} = 0 \quad \text{at} \quad \bar{y} = 0$$

and

$$EI\frac{d^2 h}{d\bar{y}^2} = EI\frac{d^3 h}{d\bar{y}^3} = 0 \quad \text{at} \quad \bar{y} = 1$$

This results in an eigenvalue problem. Eigenvalues of $\lambda$ are all negative and the lowest of these provides the divergence condition. Hence,

$$\lambda_D = -6.33 = \frac{\partial \bar{C}_L}{\partial \alpha}\frac{\sin \Lambda \cos \Lambda \; \bar{c} l^3 q}{EI}$$

The only way the right-hand side of the preceding equation can be less than zero is when $\sin \Lambda < 0$ or $\Lambda < 0$.

Thus, only swept forward wings can diverge in bending without torsional deformation. Hence, swept forward wings are most susceptible to divergence than swept back wings. This is true when both bending and torsion are present.

For many years the divergence tendency of swept forward wings precluded their use. However, composite materials with favorable bending torsion coupling alleviate this divergence.

# 8

## Rolling of a Straight Wing

### Introduction

Ailerons are used for controlling the roll motion of aircraft. By deflecting the aileron down, extra lift is created on the wing. Similarly, by deflecting the aileron up, the lift on the wing is reduced. Generally, ailerons on both wings are used in combination. If only one aileron is deployed, the extra lift force developed on that wing creates a rolling moment about the longitudinal axis of the plane.

The integral equation of equilibrium is

$$\alpha(y) = \int_0^1 C^{\alpha\alpha}(y,\eta)M_y(\eta)d\eta \tag{8.1}$$

To see what $C^{\alpha\alpha}$ is, let $M_y(\eta) = \delta(\eta - \gamma)$. Then $\alpha(y) = C^{\alpha\alpha}(y,\gamma)$. Hence, $C^{\alpha\alpha}(y,\gamma)$ is twist at y due to unit moment at $\gamma$. $C^{\alpha\alpha}(y,\eta)$ is the structural influence function.

Further, $C^{\alpha\alpha}(y,\eta) = C^{\alpha\alpha}(\eta,\gamma)$. See Figure 8.1.

### Aerodynamic Forces Including Spanwise Aerodynamic Induction

Let a rolling velocity of p rad/s roll the wing up with a velocity of p · y at a distance y from the root. This in effect gives a velocity of py down to wind (Figure 8.2).

Hence, the total angle of attack can be written as

$$\alpha_T = \alpha(y) - \frac{py}{U} \tag{8.2}$$

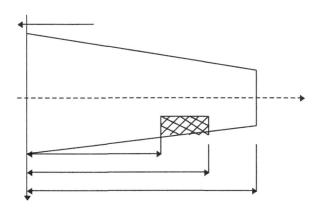

**FIGURE 8.1**
Rolling of a straight wing with ailerons.

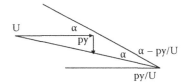

**FIGURE 8.2**
Angle of attack of a rolling wing.

The control surface is assumed rigid and its rotation is given by

$$\delta(y) = \delta_R \quad l_1 < y < l_2$$
$$= 0 \quad \text{elsewhere} \tag{8.3}$$

Let $A^{L\alpha}$ and $A^{L\delta}$ be aerodynamic influence functions, where $A^{L\alpha}$ is the lift at y due to unit $\alpha$ at $\eta$, and $A^{L\delta}$ is the lift at y due to unit $\delta$. Then,

$$C_L = \frac{L}{qc} = \frac{1}{qc}\left\{ \int_0^1 A^{L\alpha}(y,\eta)\cdot\alpha_T(\eta)d\eta + \int_0^1 A^{L\delta}(y,\eta)\delta(\eta)d\eta \right\} \tag{8.4}$$

Substituting for $\alpha_T = \alpha - py/U$ and $\delta = \delta_R\, l_1 < y < l_2$, into Equation 8.4 we get

$$C_L = \frac{1}{qc}\left\{ \int_0^1 A^{L\alpha}\cdot\alpha(\eta)d\eta - \frac{pl}{U}\int_0^1 A^{L\alpha}\frac{\eta}{l}\cdot\frac{d\eta}{l} + \delta_R\int_{l_1}^{l_2} A^{L\delta}\,\delta(\eta)d\eta \right\} \tag{8.5}$$

That is,

$$C_L = \frac{1}{qc}\left\{\int_0^1 A^{L\alpha} \cdot \alpha \, d\eta + \beta \cdot \frac{\partial C_L}{\partial \beta} + \delta_R \cdot \frac{\partial C_L}{\partial \delta_R}\right\}$$

where $\beta = py/U$.

$A^{L\delta}$ is lift coefficient at y due to unit rotation at $\delta y$.

$\frac{\partial C_L}{\partial \beta}$ is lift coefficient at y due to unit rolling velocity, $py/U \cdot \frac{\partial C_L}{\partial \beta}$ is the coefficient at y due to unit control surface rotation, $\delta_R$.

As usual,

$$C_{MAC} = \frac{M_{AC}}{qc^2} = \frac{\partial C_{MAC}}{\partial \delta_R} \cdot \delta_R \tag{8.6}$$

Note that $\frac{\partial C_{MAC}}{\partial \alpha_T} = 0$ about the aerodynamic center.

Finally, the total moment loading about the elastic axis is

$$\begin{aligned} M_y &= M_{AC} + L \cdot e \\ &= qc(C_{MAC} \cdot c + C_L \cdot e) \end{aligned} \tag{8.7}$$

Hence,

$$M_y = qc\left[c\frac{\partial C_{MAC}}{\partial \delta_R} \cdot \delta_R + e\left\{\int_0^1 A^{L\alpha} \cdot \alpha \frac{d\eta}{1} + \beta \cdot \frac{\partial C_L}{\partial \beta} + \delta_R \cdot \frac{\partial C_L}{\partial \delta_R}\right\}\right] \tag{8.8}$$

Note that $A^{L\alpha}$ and $A^{L\delta}$ are more difficult to measure than $C^{\alpha\alpha}$. It requires applying unit angles of attack at various discrete points along the span of the wing. This requires a sophisticated model and introduces experimental difficulties in establishing and maintaining a smooth flow over the airfoil. Conversely, $\frac{\partial C_L}{\partial \beta}$, $\frac{\partial C_L}{\partial \delta_R}$, $\frac{\partial C_{MAC}}{\partial \delta_R}$ are easy to measure since wings and control surfaces are considered rigid.

## Aeroelastic Equations of Equilibrium and Lumped Element Solution Method

For real-life airplane wings, Equations 8.1 and 8.8 cannot be solved exactly and we resort to a "lumped elements" approach. Equations 8.1 and 8.8 can be written as

$$\alpha(y_i) = \sum_{j=1}^{N} C^{\alpha\alpha}(y_i, \eta_j) M_y(\eta_j) \Delta\eta \quad i = 1, 2 \ldots\ldots N \tag{8.9}$$

and

$$M_y(y_i) = qc\left\{\left[c\frac{\partial C_{MAC}}{\partial \delta_R}\cdot\delta_R + e\frac{\partial C_L}{\partial\beta}\cdot\beta + e\frac{\partial C_L}{\partial\delta_R}\cdot\delta_R\right]\right.$$
$$\left. + e\sum_{j=1}^{N} A^{L\alpha}(y_i,\eta_j)\alpha(\eta_j)\frac{\Delta\eta}{1}\right\}, \ i = 1, 2 \ldots N \tag{8.10}$$

where $\Delta\eta$ is the segment width and $N$ the total number of segments.
Writing Equations 8.9 and 8.10 in matrix notation,

$$\{\alpha\} = \Delta\eta[C^{\alpha\alpha}]\{M_y\} \tag{8.11}$$

$$\{M_y\} = q[c^2]\left\{\frac{\partial C_{MAC}}{\partial\delta_R}\right\}\delta_R + q[Ce]\left\{\frac{\partial C_L}{\partial\beta}\right\}\beta$$
$$+ q[Ce]\left\{\frac{\partial C_L}{\partial\delta_R}\right\}\delta_R + q[Ce][A^{L\alpha}]\{\alpha\}\frac{\Delta\eta}{1} \tag{8.12}$$

Substituting Equation 8.12 into Equation 8.11 gives

$$\left[[1] - q\frac{(\Delta q)^2}{1}[E][A^{L\alpha}]\right]\{\alpha\} = \{f\} \tag{8.13}$$

where

$$\{f\} = q[E]\left[\left\{\frac{\partial C_L}{\partial\delta_R}\right\}\delta_R + \left\{\frac{\partial C_L}{\partial\beta}\right\}\beta\right]\Delta\eta + q[F]\left\{\frac{\partial C_{MAC}}{\partial\delta_R}\right\}\delta_R\Delta\eta$$

with

$$[E] = [C^{\alpha\alpha}][Ce]$$
$$[F] = [C^{\alpha\alpha}][C^2]$$

(8.14)

By further defining

$$[D] = \left[ [1] - q \frac{(\Delta q)^2}{1} [E][A^{L\alpha}] \right]$$

(8.15)

we can solve Equation 8.13 as

$$\{\alpha\} = [D]^{-1}\{f\}$$

(8.16)

For divergence, we have $\alpha \to \infty$ and hence

$$|D| = 0$$

(8.17)

This gives a polynomial in q. The lowest positive root (eigenvalue) gives $q_D$. The exact value of $q_D$ is obtained for $N \to \infty$. The matrix [D] does not contain $\beta$ or p. Hence, the preceding value of divergence does not depend on the rolling velocity.

In reality $\beta = pl/U$ is a function of $\delta_R$ and other problem parameters through the requirement that the wing be in static rolling equilibrium—that is, it is in equilibrium at a steady roll rate, p; the rolling moment about the x-axis is zero:

$$M_{\text{Rolling}} = 2\int_0^1 Ly \, dy = 0$$

(8.18)

In matrix notation, the discretization of Equation 8.18, $\sum_i L_i y_i \Delta y = 0$, becomes

$$2\{y\}^T\{L\}\Delta_y = 0$$

(8.19)

or

$$2q\{cy\}^T\{C_L\}\Delta_y = 0$$

From Equation 8.5 and the lumped approximation (see Equation 8.12 also),

$$\{C_L\} = \frac{\Delta\eta}{1}[A^{L\alpha}]\{\alpha\} + \left\{\frac{\partial C_L}{\partial \delta_R}\right\}\delta_R + \left\{\frac{\partial C_L}{\partial \beta}\right\}\beta \qquad (8.20)$$

Substitution of Equation 8.20 into Equation 8.19 gives

$$\{cy\}^T\left\{\frac{\Delta\eta}{1}[A^{L\alpha}]\{\alpha\} + \left\{\frac{\partial C_L}{\partial \delta_R}\right\}\delta_R + \left\{\frac{\partial C_L}{\partial \beta}\right\}\beta\right\} = 0 \qquad (8.21)$$

Equation 8.21 is a single algebraic equation. Equations 8.21 and 8.13 give N number of unknown alphas and one p value. This implies that we have N + 1 equations in N + 1 unknowns. Setting the determinant of coefficients to zero and determining the smallest positive eigenvalue gives the condition for divergence.

For $q < q_D$, $\beta$ and N number of alphas can be obtained from Equations 8.21 and 8.13. Since $\delta_R$ causes the rolling,

$$\beta \propto \delta_R$$

Hence, as $q \to q_D$, $\beta$ and alphas $\to \infty$. See Figure 8.3.

Another qualitatively different type of result (Figure 8.4) may sometimes occur as

$$\beta/\delta_R \to 0 \quad q \to q_R < q_D.$$

At $q = q_R$, "rolling reversal" is said to have occurred. $q_R$ is called the "reversal dynamic pressure." The basic phenomenon is the same as that encountered previously in "control surface reversal."

When the rolling equilibrium equation and p are considered, it corresponds to the case where there is no constraint on rolling.

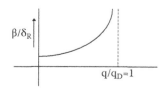

**FIGURE 8.3**
Condition for divergence.

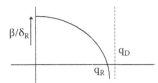

**FIGURE 8.4**
Rolling reversal.

## Determination of Aerodynamic Influence Functions

Structural influence functions are easy to determine by applying moments at various points. Aerodynamic coefficients cannot be determined that way since giving $\alpha$ at certain points may only disturb the flow over airfoil. Hence, we explain $A^{L\alpha}$ in terms of mode shapes

$$A^{L\alpha}(y,\eta) = \sum_n A_n^{L\alpha}(y)\alpha_n(n) \qquad (8.22)$$

This involves twist distributions throughout the span.

# 9

## Flutter of a Cantilever Wing

### Introduction

In Chapter 6, the flutter phenomenon was treated in a very simple fashion, considering the two degrees of freedom model for the wing. However, an aircraft wing is a continuous system with distributed mass and elasticity; moreover, the aerodynamic loading is also distributed over the length of the wing.

In this chapter, we will develop a mathematical model considering the aircraft wing as a continuous system to illustrate the flutter phenomenon (Figure 9.1).

$h(y)$ is the deflection of the elastic axis at any point y and is considered positive downward.

$\alpha\ (y)$ is the torsional deflection of the wing and is considered positive nose up.

Chordwise displacement is neglected.

Let $x_\alpha$ be distance between center of mass and elastic axis, positive if the former lies behind the latter.

In a steady flow of speed, U, the wing has some mean elastic deformation, which is of no concern in this study. The free motion of the wing about the mean position following an initial disturbance is considered.

Equations of motions include

$$\frac{\partial^2}{\partial y^2}\left(EI\frac{\partial^2 h}{\partial y^2}\right)+m\frac{\partial^2 h}{\partial t^2}+mx_\alpha\frac{\partial^2 \alpha}{\partial t^2}+L=0$$

$$-\frac{\partial}{\partial y}\left(GJ\frac{\partial \alpha}{\partial y}\right)+I_\alpha\frac{d^2\alpha}{dt^2}+mx_\alpha\frac{\partial^2 h}{\partial t^2}-M=0$$

For a uniform rod,

$$[GJ\alpha'' + M = 0 \quad \text{i.e.,} \quad GJ\alpha'' + K_\alpha \cdot \alpha = 0 \quad \text{i.e.,} \quad K_\alpha \cdot \alpha = -GJ\alpha'']$$

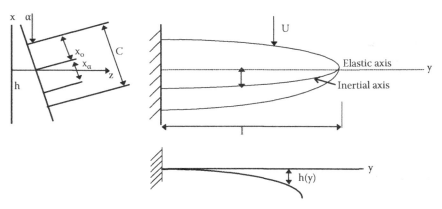

**FIGURE 9.1**
Flutter of an aircraft wing as a continuous system.

## Simple Bending of Beams

See Figure 9.2. Strain in the fiber (hatched) is

$$= \frac{(R+y)d\theta - Rd\theta}{Rd\theta}$$

$$= \frac{y}{R}$$

See Figure 9.3.

$$M = \int_{-h/2}^{h/2} \sigma \cdot y \, dy$$

$$= \int_{-h/2}^{h/2} \left(\frac{Ey}{R}\right) \cdot y \, dy$$

$$= \frac{E}{R} \int_{-h/2}^{h/2} y^2 \, dy = \frac{EI}{R}$$

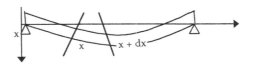

**FIGURE 9.2**
Simple bending of beam.

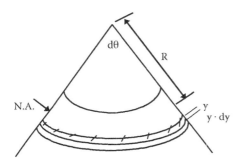

$$\therefore \frac{\text{stress}}{\text{strain}} = \frac{\sigma}{\left(y/R\right)} = E \text{ (modulus of elasticity)}$$

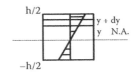

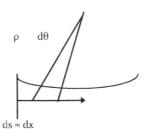

**FIGURE 9.3**
Simple bending relations.

$\rho \, d\theta = ds \approx dx$ (for small deflections):

$$\therefore \frac{d\theta}{dx} = \frac{1}{\rho} = \frac{d\left(dy/dx\right)}{dx} = \frac{d^2y}{dx^2}$$

$$\therefore M = \frac{EI}{R} = \frac{EI}{\rho} = EI \frac{d^2y}{dx^2}$$

See Figure 9.4.
Considering equilibrium of the element, we get

$$V - V - \frac{\partial V}{\partial x} dx + p dx = 0$$

$$\frac{\partial V}{\partial x} = p$$

Taking moments about the left edge, we get

$$M - M - \frac{dH}{dx} dx + \left( V + \frac{dV}{dx} dx \right) dx - p dx \cdot \frac{dx}{2} = 0$$

Considering the forces and moments acting on the element of a beam as shown in Figure 9.4.

$$\frac{dM}{dx} = V$$

We have $EIy'' = M$:

$$p = \frac{\partial V}{\partial x} = \frac{d^2 M}{dx^2} = \frac{d^2}{dx^2}(EIy'')$$

$$\frac{d^2}{dx^x}\left( EI \frac{d^2 y}{dx^2} \right) = p \qquad (9.1)$$

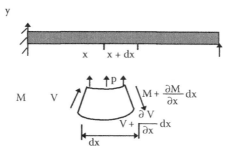

**FIGURE 9.4**
Equilibrium of a bending element.

In vibration,

$$\frac{d^2}{dx^2}\left(EI\frac{d^2y}{dx^2}\right)+M\ddot{y}=p \tag{9.2}$$

Equations 9.1 and 9.2 can be derived from the energy expressions also:

$$T=\frac{1}{2}m\int\left(\dot{h}+x_\alpha\cdot\dot{\alpha}\right)^2 dy+\frac{1}{2}I_G\int\dot{\alpha}^2\,dy \tag{9.3}$$

where $I_G$ is moment of inertia about the centroid.

$$\begin{aligned}T&=\frac{1}{2}m\int\left(\dot{h}^2+2x_\alpha\cdot\dot{h}\dot{\alpha}\right)dy+\frac{1}{2}\int\left(I_G+mx_\alpha^2\right)\dot{\alpha}^2\,dy\\&=\frac{1}{2}m\int\left(\dot{h}^2+2x_\alpha\cdot\dot{h}\dot{\alpha}\right)dy+\frac{1}{2}I_\alpha\int\dot{\alpha}^2\,dy\end{aligned}$$

where $I_\alpha$ is about the elastic axis.

$$U=\frac{1}{2}EI\int h''^2\,dy+\frac{1}{2}GJ\int\alpha'^2\,dy \tag{9.4}$$

The equations of motion are obtained by

$$\frac{d}{dt}\left(\frac{\partial T}{\partial\dot{h}}\right)+\frac{d^2}{dy^2}\left(\frac{dU}{dh''}\right)=Q_h \tag{9.5}$$

$$\frac{d}{dt}\left(\frac{\partial T}{\partial\dot{\alpha}}\right)-\frac{d}{dy}\left(\frac{dU}{d\alpha'}\right)=Q_\alpha \tag{9.6}$$

where $Q_h=-L$ and $Q_\alpha=M$. This will lead to Equations 9.1 and 9.2. Boundary conditions are

$$h=\frac{dh}{dy}=\alpha=0\quad\text{at}\quad y=0$$

$$h''=h'''=\alpha'=0\quad\text{at}\quad y=1 \tag{9.7}$$

Assume that

$$h = Af(y)e^{\lambda t} \quad \alpha = B\phi(y)e^{\lambda t} \tag{9.8}$$

where $\lambda = p + i\,\omega$, and if $p < 0$, the system is stable; if $p > 0$, it is unstable and if $p = 0$, it is just stable. This is where the onset of instability exists. Consider the solution in the form

$$h = Af(y)e^{i\omega t} \quad \alpha = B\phi(y)e^{i\omega t} \tag{9.9}$$

where $\omega$ is real and A and B are complex constants. Substituting into Equations 9.1 and 9.2, we get (assuming $\alpha_0 = 0$ and $C_{MAc0} = 0$)

$$A\left[(EIf'')'' - m\omega^2 f + i\omega\frac{k*}{U}f\right] + B[-mx_\alpha\omega^2\phi + k*\phi] = 0 \tag{9.10}$$

$$A\left[-mx_\alpha\omega^2 f - i\omega\frac{k*e}{U}f\right] + B\left[-(GJ\phi')' - I_\alpha\omega^2\phi - k*e\phi\right] = 0 \tag{9.11}$$

where $L = k*\left(\alpha + \dot{h}\Big/U\right)$, $M = Le + qc^2 C_{MAc} \cdot k* = qc\dfrac{\partial c_L}{\partial \alpha}$.

At this stage, it will be simpler to refer to Galerkin's technique. Assume $f(y)$ and $\phi(y)$ to satisfy all boundary conditions. Multiply Equation 9.10 by $f(y)dy$ and Equation 9.11 by $\phi(y)dy$ and integrate from 0 to 1. This results in

$$A(a_{11} + c_{11}\omega^2 + i\omega d_{11}) + B(c_{12}\omega^2 + i\omega d_{12}) = 0 \tag{9.12}$$

$$A(c_{21}\omega^2 + i\omega d_{21}) + B(a_{22} + c_{22}\omega^2 + d_{22}) = 0 \tag{9.13}$$

where

$$a_{11} = \int_0^1 (EIf'')''f\,dy$$

$$a_{22} = \int_0^1 -(GJ\phi')'\phi\,dy$$

$$b_{22} = -k*e\int_0^1 \phi^2\,dy$$

$$c_{11} = \int_0^1 mf^2\,dy$$

$$c_{12} = c_{21} = \int_0^1 -mx_\alpha f\phi\,dy$$

$$c_{22} = -\int_0^1 I_\alpha \phi^2 \, dy$$

$$d_{12} = k * \int_0^1 f\phi \, dy$$

$$d_{21} = k * e \int_0^1 f\phi \, dy$$

$$d_{11} = \frac{k *}{U} \int_0^1 f^2 \, dy$$

Let the frequency equation be given by getting the determinant of the coefficients to zero.

## Stability of the Motion

To study the stability of the motion, we can assume the following:

$$h = Af(y)e^{\lambda t} \qquad \alpha(y) = B\phi(y)e^{i\lambda t} \tag{9.14}$$

Substituting this into Equations 9.12 and 9.13, the characteristic equation is obtained in the form of

$$a_0\lambda^4 + a_1\lambda^3 + a_2\lambda^2 + a_3\lambda + a_4 = 0$$

The nature of the roots of a polynomial equation of this type can be studied by applying Routh's stability criterion. The method is suitable to apply on characteristic equations of control systems or physical systems. The polynomial equation must contain only a finite number of terms. The coefficients of the characteristic equations are arranged in a specific formation and the method will provide information on the positive roots of the characteristic equation that would indicate that the system is unstable.

Given a polynomial equation of the type

$$a_0 s^n + a_1 s^{n-1} + \ldots + a_{n-1}s + a_n = 0 \tag{9.15}$$

the coefficient $a_n$ must be nonzero to ensure that $s = 0$ is not a root.

The presence of a zero as a coefficient or a negative coefficient along with positive coefficients indicates right away a root or roots that may be imaginary in nature or have positive real parts. This situation indicates that the system is unstable.

The polynomial with real coefficients can be factored into either linear factors of the type $(s + a)$ or quadratic factors such as $(s^2 + bs + c)$, where a, b, and c are real. The linear factors indicate real roots while the quadratic factors may have real or complex conjugate pairs of roots. The quadratic factor $(s^2 + bs + c)$ has complex roots with negative real parts if b and c are both positive. If a, b, c, etc. in all factors are positive, then the roots of the polynomial are negative or they are complex with negative real parts. Hence, all coefficients, $a_i$, being positive is a necessary condition for stability. However, they are not sufficient to ensure stability.

The sufficient condition for stability is obtained by the following procedure: Arrange the coefficients, $a_i$, of the polynomial in the following formation in the form of rows and columns:

| | | | | | |
|---|---|---|---|---|---|
| $s^n$: | $a_0$ | $a_2$ | $a_4$ | $a_6$ | ... |
| $s^{n-1}$: | $a_1$ | $a_3$ | $a_5$ | $a_7$ | ... |
| $s^{n-2}$: | $b_1$ | $b_2$ | $b_3$ | $b_4$ | ... |
| $s^{n-3}$: | $c_1$ | $c_2$ | $c_3$ | $c_4$ | ... |
| $s^{n-4}$: | $d_1$ | $d_2$ | $d_3$ | $d_4$ | ... |
| ⋮ | | | | | |
| $s^3$: | $e_1$ | $e_2$ | $e_3$ | | |
| $s^2$: | $f_1$ | $f_2$ | | | |
| $s^1$: | $g_1$ | 0 | | | |
| $s^0$: | $h_1$ | | | | |

where

$$b_1 = \frac{a_1 a_2 - a_0 a_3}{a_1}$$

$$b_2 = \frac{a_1 a_4 - a_0 a_5}{a_1}$$

$$b_3 = \frac{a_1 a_6 - a_0 a_7}{a_1} \quad \text{etc.}$$

$$c_1 = \frac{b_1 a_3 - a_1 b_2}{b_1}$$

$$c_2 = \frac{b_1 a_5 - a_1 b_3}{b_1}$$

$$c_3 = \frac{b_1 a_7 - a_1 b_4}{b_1} \quad \text{etc.}$$

$$d_1 = \frac{c_1 b_2 - b_1 c_2}{c_1}$$

$$d_2 = \frac{c_1 b_3 - b_1 c_3}{c_1}$$

$$d_3 = \frac{c_1 b_4 - b_1 c_4}{c_1} \quad \text{etc.}$$

$$g_1 = \frac{f_1 e_2 - e_1 f_2}{f_1}$$

$$h_1 = \frac{g_1 f_2 - 0}{g_1} \quad \text{etc.}$$

The array of coefficients $a_i$, $b_i$, $c_i$, $d_i$, etc. assumes a triangular form.

According to Routh's stability criterion, the number of sign changes in the first column (i.e., $a_0$, $a_1$, $b_1$, $c_1$, ... $f_1$, $g_1$, $h_1$) indicate the number of roots with positive real parts.

Hence, the necessary and sufficient condition for stability of a system is that its characteristic equation has all positive coefficients and that all the coefficients in the first column of the array as formed have positive signs.

### Example 9.1

Consider the stability of a system with a characteristic equation given by

$$s^5 + 3s^4 + 2s^3 + s^2 + 4s + 3 = 0$$

Note that all the coefficients are positive. Hence, the necessary condition for stability is satisfied. Arrange the coefficients and construct the triangular formation as follows:

| | | | |
|---|---|---|---|
| $s^5$ | 1 | 2 | 4 |
| $s^4$ | 3 | 1 | 3 |
| $s^3$ | 5/3 | 3 | 0 |
| $s^2$ | −4.4 | 3 | 0 |
| $s^1$ | 4.14 | 0 | |
| $s^0$ | 3 | | |

In the fourth row, there is one sign change and the sign changes again
in the fifth row in the first column. Hence, there are two sign changes,
indicating two roots with positive parts. The system is unstable. Note
that wherever coefficients are missing, they may be taken as zero in the
array.

**Example 9.2**

Check the stability of a system with the following characteristic equation:

$$s^3 + 2s^2 + 2s + 4 = 0$$

Note that all the coefficients are positive, satisfying the necessary condition for stability. Further,

$$
\begin{array}{llll}
s^3 & 1 & 2 & 0 \\
s^2 & 2 & 4 & \\
s^1 & 0 \approx \varepsilon & 0 & \\
s^0 & 4 & &
\end{array}
$$

Note that the zero coefficient in the first column of the third row is
replaced by $\varepsilon$, a small positive value, and the array is continued using
this value. If the signs of the coefficients above and below the zero coefficient are the same, it indicates the presence of a pair of imaginary roots.
The system is stable.

However, if the signs above and below the zero are opposite, it indicates one sign change.

Considering the flutter problem of the two degrees of freedom system
resulting in a fourth-order characteristic equation, for the real parts of
the roots to be negative, the Hurwitz determinants must all be positive.
Arranging the coefficients as the following,

$$
\begin{array}{llll}
\lambda^4 & a_0 & a_2 & a_4 \\
\lambda^3 & a_1 & a_3 & 0 \\
\lambda^2 & b_1 & b_2 & b_3 \\
\lambda' & c_1 & c_2 & c_3 \\
\lambda^0 & d_1 & &
\end{array}
$$

$$b_1 = (a_1a_2 - a_0a_3)/a_1$$
$$b_2 = (a_1a_4 - a_0a_5)/a_1 = a_4$$
$$c_1 = (b_1a_3 - a_1b_2)/b_1, \quad c_2 = (b_1a_5 - a_1b_3)/b_1$$

Necessary and sufficient conditions for stability and $a_0$, $a_1$, $a_2$, $a_3$, $a_4$, $b_1$, $c_1$, etc. are positive. $c_1 > 0$—that is,

$$\frac{\left(\dfrac{a_1 a_2 - a_0 a_3}{a_1}\right) a_3 - a_1 a_4}{b_1} > 0$$

That is, $a_1 a_2 a_3 - a_0 a_3^2 - a_1^2 a_4 > 0$.

# 10

## Approximate Techniques of Modeling Continuous Systems

### Influence Coefficient Method

The influence coefficient defines a certain output due to a specific input. The influence coefficient in Figure 10.1 defines the deflection at a certain point on the structure due to a unit load input at another location. The quantity $C^{y\eta}(y,\eta)$ defines the deflection at y due to unit load at $\eta$.

Using this influence coefficient, it is possible to obtain the deflection at any point due to all the loads acting on the structure as

$$w(y) = \int_0^1 C(y,\eta)p(\eta)d\eta$$

Discretizing this integral using two points at 1 and 2, we get

$$w(y_i) = \sum_{j=1}^2 C(y_i, \eta_i)\, p(\eta_j)\Delta\eta \quad i = 1,2 \tag{10.1}$$

Let $\Delta\eta = 1/2$, $y_1 = 1/2$, and $y_2 = 1$:

$$w(y_1) = C(y_1, \eta_1)p(\eta_1)1/2 + C(y_1, \eta_2)p(\eta_2)1/2$$
$$w(y_2) = C(y_2, \eta_1)p(\eta_1)1/2 + C(y_2, \eta_2)p(\eta_2)1/2$$

For vibration problems,

$$p(\eta) = m\omega^2 w(\eta)$$

**FIGURE 10.1**
Influence coefficients.

Hence,

$$w(y_1) = C(y_1, \eta_1)\, m\omega^2\, w(\eta_1)1/2 + C(y_1, \eta_2)\, m\omega^2 w(\eta_2)\, 1/2$$

$$w(y_2) = C(y_2, \eta_1)\, m\omega^2 w(\eta_1)1/2 + C(y_2, \eta_2)\, m\omega^2 w(\eta_2)\, 1/2$$

$$ml/2 = M/2 \tag{10.2}$$

Hence,

$$\begin{bmatrix} \left(C_{11}\,\dfrac{M}{2} - \dfrac{1}{\omega^2}\right) & C_{12}\,\dfrac{M}{2} \\[3mm] C_{21}\,\dfrac{M}{2} & C_{22}\,\dfrac{M}{2} - \dfrac{1}{\omega^2} \end{bmatrix} \begin{Bmatrix} w_1 \\ w_2 \end{Bmatrix} = \begin{Bmatrix} 0 \\ 0 \end{Bmatrix} \tag{10.3}$$

$C_{ij}$ can be obtained from static deflection considerations:

$$y = \frac{P}{2EI}\left(lx^2 - \frac{x^3}{3}\right)$$

## Galerkin's Method

See Figure 10.2.
  The equation of motion is

$$EIw'''' + m\ddot{w} = 0 \tag{10.4}$$

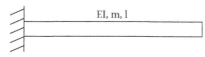

**FIGURE 10.2**
Galerkin's method.

Assume that

$$w(x) = \sum_{i=1}^{N} a_i f_i(x)$$

where $f_i(x)$ satisfy all the boundary conditions and $a_i$ are arbitrary constants. Consider a virtual displacement:

$$\delta w = \sum_{i=1}^{N} \delta\, a_i f_i(x)$$

The corresponding virtual work is obtained as

$$\delta W = \int_0^1 [EIw''' + m\ddot{w}]\delta w\, dx$$

$$= \int_0^1 \left( \sum_{i=1}^{N} EI a_i f_i''' + m\ddot{a}_i f_i \right) \sum_{j=1}^{N} \delta a_j f_j(x)\, dx$$

Let

$$a_i = a_i\, e^{i\omega t} \tag{10.5}$$

Assuming that the virtual work is invariant with respect to the arbitrary coefficients, $a_k$, we have

$$\frac{\delta W}{\delta a_k} = \int_0^1 \sum a_i \left[ EIf_i''' - m\omega^2 f_i \right] f_k(x)\, dx = 0 \tag{10.6}$$

For every k, there is an equation and hence we have N simultaneous equations in N unknowns.

Consider a one-term solution for a cantilever beam, satisfying all the boundary conditions. Such a solution is obtained as follows:

Let

$$f_1(x) = a_0 + a_1 x + a_2 x^2 + a_3 x^3 + a_4 x^4$$

Applying boundary conditions, $f_1(0) = f_1'(0) = 0$.

Hence, $a_0 = a_1 = 0$
Again, $f''(1) = f'''(1) = 0$; that is,

$$2a_2 + 6a_3 + 12a_4 = 0$$
$$6a_3 + 24a_4 = 0$$
$$a_3 = -4a_4$$
$$a_2 = 6a_4$$

$$f_1(x) = 6x^2 - 4x^3 + x^4$$

Hence,

$$\int_0^1 [EI\,24 - m\omega^2(6x^2 - 4x^3 + x^4)](6x^2 - 4x^3 + 4x^4)\,dx = 0$$

That is,

$$24EI\left(2 - 1 + \tfrac{1}{5}\right) - m\omega^2\left(\frac{36}{5} + \frac{16}{7} + \frac{1}{9} - \frac{48}{6} + \frac{12}{7} - \frac{8}{8}\right)$$

That is,

$$\omega^2 = (28.8/2.311)(EI/m) = 12.6EI/m$$

Hence,

$$\omega = 3.53\sqrt{EI/m}$$

Exact value = 3.516 EI/m.

## Rayleigh-Ritz Method

The Rayleigh-Ritz method is used to obtain the approximate natural frequencies and the natural modes of a structure. In this method, an approximate expression for the deflection of the structure is assumed that satisfies at least the geometric boundary conditions of the structure. This approximate deflection is used to obtain the maximum values of the potential and kinetic energies. In a conservative system, the maximum potential energy must equal the maximum kinetic energy.

In the Rayleigh-Ritz method, the approximate deflection expression is assumed in the form of a linear combination of several expressions, each of which satisfies at least the geometric boundary conditions of the structure. The energy expressions are calculated and an expression for the natural frequency is obtained. This natural frequency is made stationary with respect to the arbitrary coefficients appearing in the deflection expression. The method is illustrated in Figure 10.3.

Let

$$w(x) = \sum a_i f_i(x)$$

where $f_i(x)$ need satisfy only the geometrical boundary conditions. The maximum kinetic and potential energy expressions are given, respectively, by

$$T_{max} = \frac{1}{2} m\omega^2 \int_0^1 w^2 \, dx = \omega^2 T_{max}^*$$

$$U_{max} = \frac{1}{2} EI \int w''^2 \, dx$$

$$T_{max} = U_{max}$$

Since $T_{max} = \omega^2 T_{max}^*$, we can write

$$\omega^2 = \frac{U_{max}}{T_{max}^*}$$

**FIGURE 10.3**
The Rayleigh Ritz method.

The natural frequency is stationary with respect to the arbitrary coefficient $a_j$ and hence we get

$$\frac{\partial \omega^2}{\partial a_j} = \frac{T^*_{max}\dfrac{\partial U_{max}}{\partial a_j} - U_{max}\dfrac{\partial T^*_{max}}{\partial a_j}}{T^*_{max}} = 0$$

That is,

$$\frac{\partial U_{max}}{\partial a_j} - \omega^2 \frac{\partial T^*_{max}}{\partial a_j} = 0$$

$$U_{max} = \frac{EI}{2}\int_0^1 \left(\sum_{i=1}^N a_i f_i''\right)^2 dx$$

$$\frac{\partial U}{\partial a_j} = EI \int_0^1 \sum_{i=1}^N a_i f_i'' f_j'' dx$$

$$\frac{\partial T^*_{max}}{\partial a_j} = m\omega^2 \int_0^1 \sum_{i=1}^N a_i f_i f_j \, dx$$

Thus, we obtain the following eigenvalue problem:

$$([A] - \omega^2[B])\{x\} = \{0\} \tag{10.7}$$

Consider

$$\omega = a_1 \left(\frac{x}{1}\right)^2 + a_2 \left(\frac{x}{1}\right)^3$$

Both satisfy geometrical boundary conditions for cantilever.

$$[B] = \begin{bmatrix} \frac{1}{5} & \frac{1}{6} \\ \frac{1}{6} & \frac{1}{7} \end{bmatrix} \quad [A] = \begin{bmatrix} 4 & 6 \\ 6 & 12 \end{bmatrix}$$

That is,

$$\begin{vmatrix} \left(4 - \frac{1}{5}\omega^2\right) & \left(6 - \frac{1}{6}\omega^2\right) \\ \left(6 - \frac{1}{6}\omega^2\right) & \left(12 - \frac{1}{7}\omega^2\right) \end{vmatrix} = 0$$

$$48 - \omega^2\left(\frac{12}{5} + \frac{4}{7}\right) + \frac{\omega^4}{35} - \left(36 + \frac{\omega^4}{36} - 2\omega^2\right) = 0$$

That is,

$$0.00079365 1\omega^2 - 0.97143\omega^2 + 12 = 0$$

That is,

$$\omega^4 - 1224\omega^2 + 15120 = 0$$

$$\omega^2 = \frac{1224 \pm 1199}{2}$$

$$= 12.5 \text{ or } 1211.5$$

$$\omega = 3.536\sqrt{\frac{EI}{ml^2}} \quad \text{or} \quad 34.81\sqrt{\frac{EI}{ml^2}}.$$

## Finite Element Method

Consider a beam element undergoing bending deflection. Consider two degrees of freedom at each node—say, w and θ—corresponding to deflection and slope. We can express the deflection (Figure 10.4) in general as

$$w(x) = a_0 + a_1 x + a_2 x^2 + a_3 x^3 \tag{10.8}$$

$$w_1 = a_0$$

$$\theta_1 = a_1$$

**FIGURE 10.4**
Finite element.

$$w_2 = a_0 + a_1 + a_2 + a_3$$

$$\theta_2 = a_1 + 2a_2 + 3a_3$$

$$\{q\} = \begin{bmatrix} 1 & 0 & 0 & 0 \\ 0 & 1 & 0 & 0 \\ 1 & 1 & 1 & 1 \\ 0 & 1 & 2 & 3 \end{bmatrix} \begin{Bmatrix} a_0 \\ a_1 \\ a_2 \\ a_3 \end{Bmatrix}$$ (10.9)

$$\therefore \{a\} = A^{-1}\{q\}$$

$$w(x) = \begin{pmatrix} 1 & x & x^2 & x^3 \end{pmatrix} \{a\}$$

$$= \begin{pmatrix} 1 & x & x^2 & x^3 \end{pmatrix} A^{-1}\{q\} = X^T A^{-1} q$$

$$T = \frac{1}{2} m\omega^2 l \int_0^1 w^2 \, dx$$

$$= \frac{1}{2} m\omega^2 l \int_0^1 q^T (A^{-1})^T X \cdot X^T A^{-1} q \, dx$$ (10.10)

$$= \frac{1}{2} \omega^2 q^T [M_e] q$$

where

$$M_e = m(A^{-1})^T l \int_0^1 (XX^T) \, dx A^{-1}$$

is the mass matrix. We have

$$
XX^T = \begin{bmatrix} 1 \\ x \\ x^2 \\ x^3 \end{bmatrix} \begin{pmatrix} 1 & x & x^2 & x^3 \end{pmatrix}
$$

$$
= \begin{bmatrix} 1 & x & x^2 & x^3 \\ x & x^2 & x^3 & x^4 \\ x^2 & x^3 & x^4 & x^5 \\ x^3 & x^4 & x^5 & x^6 \end{bmatrix}
$$

(10.11)

$$
U = \frac{1}{2}\frac{EI}{l^3} \int w''^2 \, dx
$$

$$
= \frac{1}{2}\frac{EI}{l^3} \int q^T (A^{-1})^T X'' X''^T A^{-1} q \, dx
$$

$$
= \frac{1}{2} q^T [K] q
$$

(10.12)

$$
[E] = \frac{EI}{l^3} (A^{-1})^T \left( \int X'' X''^T \, dx \right) (A^{-1})
$$

is the stiffness matrix function,

$$
X'' X''^T = \begin{Bmatrix} 0 \\ 0 \\ 2 \\ 6x \end{Bmatrix} \begin{pmatrix} 0 & 0 & 2 & 6x \end{pmatrix}
$$

$$
= \begin{bmatrix} 0 & 0 & 0 & 0 \\ 0 & 0 & 0 & 0 \\ 0 & 0 & 4 & 12x \\ 0 & 0 & 12x & 36x^2 \end{bmatrix}
$$

## Assembly of Finite Elements

Since the element coordinates are aligned with the global coordinates of the assembly, we can keep the individual elements side by side and consider the degrees of freedom at the common nodes as common degrees of freedom. This ensures connectivity in deflections and slopes there accordingly (Figure 10.5),

$$
m_e^{(1)} = \begin{bmatrix}
m_{11}^{(1)} & m_{12}^{(1)} & m_{13}^{(1)} & m_{14}^{(1)} \\
m_{21}^{(1)} & m_{22}^{(1)} & m_{23}^{(1)} & m_{24}^{(1)} \\
m_{31}^{(1)} & m_{32}^{(1)} & m_{33}^{(1)} & m_{34}^{(1)} \\
m_{41}^{(1)} & m_{42}^{(1)} & m_{43}^{(1)} & m_{44}^{(1)}
\end{bmatrix}
$$

$$
M = \begin{bmatrix}
m_{11}^{(1)} & m_{12}^{(1)} & m_{13}^{(1)} & m_{14}^{(1)} & 0 & 0 \\
 & & m_{23}^{(1)} & m_{24}^{(1)} & 0 & 0 \\
 & & m_{33}^{(1)} + m_{11}^{(2)} & m_{34}^{(1)} + m_{12}^{(2)} & m_{13}^{(2)} & m_{14}^{(2)} \\
 & & m_{13}^{(1)} + m_{21}^{(2)} & m_{14}^{(1)} + m_{22}^{(2)} & m_{23}^{(2)} & m_{24}^{(2)} \\
 & & & & m_{33}^{(2)} & m_{34}^{(2)} \\
 & & & & & m_{44}^{(2)}
\end{bmatrix}
$$

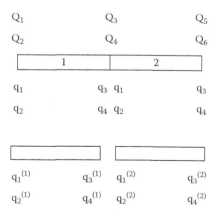

FIGURE 10.5
Assembly of finite elements.

$$m_e^{(2)} = \begin{bmatrix} m_{11}^{(2)} & m_{12}^{(2)} & m_{13}^{(2)} & m_{14}^{(2)} \\ m_{21}^{(2)} & m_{22}^{(2)} & m_{23}^{(2)} & m_{24}^{(2)} \\ m_{31}^{(2)} & m_{32}^{(2)} & m_{33}^{(2)} & m_{34}^{(2)} \\ m_{41}^{(2)} & m_{42}^{(2)} & m_{43}^{(2)} & m_{44}^{(2)} \end{bmatrix}$$

**Example 10.1**

For a beam,

$$[m_e] = \frac{\rho A L}{420} \begin{bmatrix} 156 & 22L & 54 & -13L \\ 22L & 4L^2 & 13L & -3L^2 \\ 54 & 13L & 156 & -22L \\ -13L & -3L^2 & -22L & 4L^2 \end{bmatrix}$$

$$[k_e] = \frac{EI}{L^3} \begin{bmatrix} 12 & & & \text{symmetric} \\ 6L & 4L^2 & & \\ -12 & -6L & 12 & \\ 6L & 2L^2 & -6L & 4L^2 \end{bmatrix}$$

1. Consider a cantilever beam (Figure 10.6)

$$[m] = \frac{\rho A L}{420} \begin{bmatrix} 156 & -22L \\ -22L & 4L^2 \end{bmatrix} = \rho A L \begin{bmatrix} 13/35 & -11L/210 \\ -11L/210 & L^2/105 \end{bmatrix}$$

$$[k] = \frac{EI}{L^3} \begin{bmatrix} 12 & -6L \\ -6L & 4L^2 \end{bmatrix}$$

$$m\ddot{x} + kx = 0 \qquad x = Xe^{i\omega t}$$

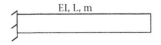
EI, L, m

**FIGURE 10.6**
Cantilever beam.

$$|k - m\omega^2| = 0 \quad \text{Let } \frac{l^3}{EI} \cdot \rho Al \cdot \omega^2 = \lambda$$

$$\begin{vmatrix} 12 - (13/35)\lambda & -6L + (11L/210)\lambda \\ -6L + (11L/210)\lambda & 4L^2 - L^2\lambda/105 \end{vmatrix} = 0$$

That is,

$$\left(12 - \frac{13}{35}\lambda\right)\left(4l^2\frac{L^2\lambda}{105}\right) - \left(6L - \frac{11L\lambda}{210}\right)^2 = 0$$

That is,

$$48 - \left(\frac{12}{105} + \frac{52}{35}\right)\lambda + \frac{13\lambda^2}{35 \times 105} - \left(36 - \frac{132\lambda}{210} + \frac{12l\lambda^2}{210^2}\right) = 0$$

Solving for $\lambda$, we get

$$0.000793651\lambda^2 - 0.9714286\lambda + 12 = 0$$

$$\lambda = \frac{0.9714286 \pm 0.9516188}{0.001587302} = 12.48 \text{ or } 1211.52$$

$$\omega = 3.533\sqrt{\frac{EI}{\rho AL^4}} \quad \text{or} \quad 34.81\sqrt{\frac{EI}{\rho AL^4}}$$

$$\omega_{\text{exact}} = 3.516\sqrt{\frac{EI}{\rho AL^4}} \quad \text{or} \quad 22.0345\sqrt{\frac{EI}{\rho AL^4}}$$

2. Consider two elements for a beam with both ends clamped (Figure 10.7).

L            L

**FIGURE 10.7**
A two-element beam.

Assembling the mass and stiffness matrices,

$$M = \frac{\rho AL}{420} \begin{bmatrix} 156 & 22L & 54 & -13L & 0 & 0 \\ 22L & 4L^2 & 13L & -3L^2 & 0 & 0 \\ 54 & 131 & 312 & 0 & 54 & -13L \\ -13L & -3L^2 & 0 & 8L^2 & 13L & -3L^2 \\ 0 & 0 & 54 & 13L & 156 & -22L \\ 0 & 0 & -13L & -3L^2 & -22L & 4L^2 \end{bmatrix}$$

$$K = \frac{EI}{L^3} \begin{bmatrix} 12 & 6L & -12 & 6L & 0 & 0 \\ 6L & 4L^2 & -6L & 2L^2 & 0 & 0 \\ -12 & -6L & 24 & 0 & -12 & -6L \\ 6L & 2L^2 & 0 & 8L^2 & -6L & 2L^2 \\ 0 & 0 & -12 & -6L & 12 & -6L \\ 0 & 0 & 6L & 2L^2 & -6L & 4L^2 \end{bmatrix}$$

In view of the deflection and slope being zero at the clamped ends, the modal coordinates $q_1$, $q_2$, $q_5$, and $q_6$ are zero. Consequently, eliminating 1,2 and 5,6 rows and columns,

$$M = \frac{\rho AL}{420} \begin{bmatrix} 312 & 0 \\ 0 & 8L^2 \end{bmatrix} \quad K = \begin{bmatrix} 24 & 0 \\ 0 & 8L^2 \end{bmatrix}$$

$$= \rho Al \begin{bmatrix} \dfrac{26}{35} & 0 \\ 0 & 2L^2/105 \end{bmatrix}$$

$$|k - m\lambda| = 0$$

where

$$\lambda = \omega L^2 \sqrt{\frac{\rho A}{EI}}$$

$$\begin{vmatrix} \left(24 - \dfrac{26\lambda}{35}\right) & 0 \\ 0 & \left(8L^2 - \dfrac{2L^2}{105}\lambda\right) \end{vmatrix} = 0$$

That is,

$$\lambda_1 = \frac{35 \times 24}{26} \qquad \lambda_2 = \frac{105 \times 8}{2}$$

$$= \frac{420}{13} = 32.3 \qquad = 420$$

$$\omega_1 = \sqrt{32.3} \times \sqrt{\frac{EI}{\rho AL^4}} \qquad \omega_2 = 20.49 \times \sqrt{\frac{EI}{\rho AL^4}}$$

$$= 5.683 \sqrt{\frac{EI}{\rho AL^4}}$$

$$\omega_{1exact} = 22.3733 \sqrt{\frac{EI}{\rho AL^4}} \qquad \omega_{2exact} = 61.6728 \sqrt{\frac{EI}{\rho AL^4}}$$

Hence, taking only two elements for this case results in a very poor approximation for the beam with both ends clamped.

# Finite Element Representation of the Response and Flutter Problems

## Beam-Rod Model

The kinetic and potential energy expressions are

$$T = \frac{1}{2} m \int_0^1 \left(\dot{h} + x_\alpha \cdot \dot{\alpha}\right)^2 dy + \frac{1}{2} I_G \int \dot{\alpha}^2 dy$$

$$= \frac{1}{2} m \int_0^1 \left(\dot{h}^2 + 2x_\alpha \dot{h}\dot{\alpha}\right) dy + \frac{1}{2} I_\alpha \int \dot{\alpha}^2 dy$$

$$U = \frac{1}{2} EI \int h''^2 \, dy + \frac{1}{2} GJ \int \alpha'^2 \, dy$$

In Figure 10.8, the element has six degrees of freedom, where $h_1$ and $h_2$ denote deflections, $\theta_1$ and $\theta_2$ denote slope, and $\alpha_1$ and $\alpha_2$ denote torsional twists. Let

$$h(y) = a_0 + a_1 y + a_2 y^2 + a_4 y^3 = \begin{pmatrix} 1 & y & y^2 & y^3 \end{pmatrix} \{a\}$$

$$\alpha(y) = a_5 + a_6 y = \begin{pmatrix} 1 & y \end{pmatrix} \{a\}$$

$$\{q\} = \begin{Bmatrix} h_1 \\ \theta_1 \\ \alpha_1 \\ h_2 \\ \theta_2 \\ \alpha_2 \end{Bmatrix} = \begin{bmatrix} 1 & 0 & 0 & 0 & 0 & 0 \\ 0 & 1 & 0 & 0 & 0 & 0 \\ 0 & 0 & 0 & 0 & 1 & 0 \\ 1 & L & L^2 & L^3 & 0 & 0 \\ 0 & 1 & 2L & 3L^2 & 0 & 0 \\ 0 & 0 & 0 & 0 & 1 & L \end{bmatrix} \{a\}$$

This is symbolically written as

$$\{q\} = A\{\alpha\}$$

Hence,

$$\{a\} = A^{-1} q$$

$$h(y) = \begin{pmatrix} 1 & y & y^2 & y^3 & 0 & 0 \end{pmatrix} \{a\} = Y_h \cdot \{a\} = Y_h A^{-1} \{q\}$$

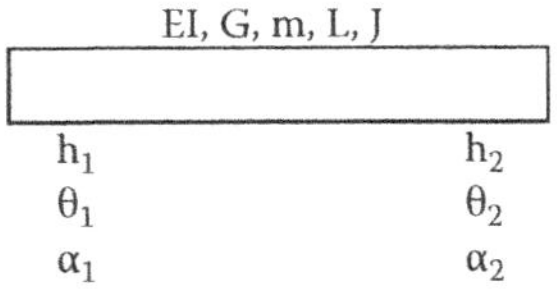

FIGURE 10.8
Beam-rod model.

$$\alpha(y) = \begin{pmatrix} 0 & 0 & 0 & 0 & 1 & y \end{pmatrix}\{a\} = Y_\alpha A^{-1}q$$

$$T = \frac{1}{2}m\omega^2 \int_0^1 q^T A^{-1T} Y_h Y_h^T A^{-1}q\,dy$$

$$+ \frac{1}{2}m\omega^2 \int_0^1 x_\alpha(y)q^T A^{-1T}\left(Y_h Y_\alpha^T + Y_\alpha Y_h^T\right)A^{-1}q\,dy$$

$$+ \frac{1}{2}I_\alpha\omega^2 \int_0^1 q^T A^{-1T} Y_\alpha . Y_\alpha^T A^{-1}q\,dy$$

$$= \frac{1}{2}\omega^2 q^T [M_e]q$$

Similarly,

$$U_s = \frac{1}{2}EI \int_0^1 q^T A^{-1} Y_h'' Y_h''^T A^{-1}q\,dy$$

$$+ \frac{1}{2}GJ \int_0^1 q^T A^{-1} Y_\alpha' Y_\alpha'^T A^{-1}q\,dy$$

$$= \frac{1}{2}q^T [K_e]q$$

Lift force does work due to h motion as well as $\alpha$ motion. Hence, finite elements must be constructed to account for these terms also:

$$U_L = -\frac{1}{2}L.h + \frac{1}{2}M_y \cdot \alpha$$

$$= -\frac{1}{2}qc\frac{\partial C_L}{\partial \alpha} \cdot \alpha \cdot h + \frac{1}{2}e\left(q_c \frac{\partial C_L}{\partial \alpha} \cdot \alpha\right)\alpha$$

$$= -\frac{1}{2}K^*\alpha \cdot h + \frac{1}{2}K^*e \cdot \alpha^2$$

This gives an "aerodynamic stiffness" element $[K_{ae}]$ as well as an "aerodynamic damping" element.

Finally,

$$\left[\alpha_T = \alpha + \frac{\dot{h}}{U}\right]$$

The equations of motion are given by

$$[M_e]\{\ddot{q}\} + [C_e]\{\dot{q}\} + [K_e + K_{ae}]\{q\} = \{F\}$$

where {F} is due to any additional gust loading.

Dynamic response problems can be solved by considering

$$M_e\ddot{q} + K_e q = f(t)$$

where {f(t)} includes forces due to lift also. Normal modes and natural frequencies can be obtained and the response can be evaluated in terms of the normal modes.

By solving the complex eigenvalue problem, we can solve the flutter problem.

For simple bending,

$$EIw'''' + m\ddot{w} = L(x)$$

Consider free harmonic vibration. Then,

$$EIw'''' + m\ddot{w} = 0$$

with

$$w(x,t) = W(x)e^{i\omega t}$$

and

$$EIW'''' - m(\omega)^2 W = 0$$

Let $k^4 = m\omega^2/EI$ and hence

$$W'''' - k^4W = 0$$

Assume $W = A\,e^{i\omega t}$. Then,

$$\lambda^4 - k^4 = 0.$$

Hence,

$$\lambda = \pm k \quad \text{and} \quad \pm ik$$

Hence, the solution can be expressed as

$$W(x) = A_1e^{\lambda x} + A_2e^{-\lambda x} + A_3e^{i\lambda x} + A_4e^{-i\lambda x} \tag{10.13}$$

or it can be expressed in the form of

$$W(x) = C_1\cosh\lambda x + C_2\sinh\lambda x + C_3\cos\lambda x + C_4\sin\lambda x$$

### Example 10.2: A Cantilever Beam

Boundary conditions are

$$W(0) = W'(0) = W''(L) = W'''(L) = 0$$

Applying boundary conditions to the solution, we obtain

$$C_1 + C_3 = 0 \quad C_3 = -C_1$$
$$C_2 + C_4 = 0 \quad C_4 = -C_2$$

Hence,

$$W(x) = C_1(\cosh\lambda x - \cos\lambda x) + C_2(\sinh\lambda x - \sin\lambda x)$$

$$W''(x) = \lambda^2[C_1(\cosh\lambda x + \cos\lambda x) + C_2(\sinh\lambda x + \sin\lambda x)]$$

$$W'''(x) = \lambda^3[C_1(\sinh\lambda x - \sin\lambda x) + C_2(\cosh\lambda x + \cos\lambda x)]$$

Applying the last two boundary conditions,

$$[C_1(\cosh\lambda l + \cosh\lambda l) + C_2(\sinh\lambda l + \sin\lambda l)] = 0$$

$$[C_1(\sinh\lambda l - \sin\lambda l) + C_2(\cosh\lambda l + \cosh\lambda l] = 0$$

From this, the frequency equation is obtained as

$$(\cosh\lambda l + \cos\lambda l)^2 - (\sinh^2\lambda l - \sin^2\lambda l) = 0$$

That is,

$$\cosh^2\lambda l - \sinh^2\lambda l + \cos^2\lambda l + \sin^2\lambda l + 2\cosh\lambda l \cdot \cosh\lambda l = 0$$

$$2 + 2\cosh\lambda l \cdot \cos\lambda l = 0$$

or

$$\cosh\lambda l \cdot \cos\lambda l + 1 = 0$$

See Figure 10.9.

$$\lambda^4 = \frac{m\omega^2}{EI} \quad (\lambda l)^4 = \frac{m\omega^2 l^4}{EI}$$

$$\omega^2 = (\lambda l)^4 \sqrt{\frac{EI}{ml^4}} = 3.516 \sqrt{\frac{EI}{ml^4}}$$

and

$$22.0345 \sqrt{\frac{EI}{ml^4}}$$

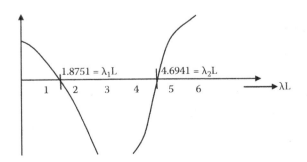

**FIGURE 10.9**
Eigenvalues of a cantilever beam.

**FIGURE 10.10**
Beam on elastic foundation.

## Elastic Foundation

The equation of motion for a beam resting on an elastic foundation with a distributed stiffness coefficient of k per unit length (Figure 10.10) is

$$EIW'''' + kW - m\omega^2 W = 0$$

That is,

$$W'''' + \left( \frac{k - m\omega^2}{EI} \right) W = 0$$

Let

$$-\frac{-k + m\omega^2}{EI} = \lambda^4$$

Then, following similar analysis as before, the natural frequencies are obtained as $\lambda_1, \lambda_2$, etc. Then,

$$\frac{m\omega^2}{EI} = \lambda^4 - \frac{k}{EI} \therefore \omega^2 = \lambda^4 l^4 \cdot \frac{EI}{ml^4} - \frac{k}{m}$$

## Eigenfunctions and Eigenvalues

Returning to the discussion on the frequency equation of beams, the eigenvalues, when substituted into the solution, provide the eigenfunctions. Let $\omega_1, \omega_2$, etc., corresponding to $\lambda_1, \lambda_2$, etc., be the natural frequencies. Corresponding W(x) are the eigenfunctions $\phi_n(x)$. As in the case of torsional vibrations, it is possible to show that the $\phi_n(x)$ are orthogonal to each other. The equation of motion is

$$W'''' - \lambda^4 W = 0$$

Since the eigenfunctions are solutions, we have

$$\phi_m'''' - \lambda_m^4 \phi_m$$

$$\phi_n'''' - \lambda_n^4 \phi_n = 0$$

Multiplying the first equation by $\phi_n$ and the second by $\phi_m$ and integrating from 0 to 1, we get

$$\int_0^1 \phi_m'''' \phi_n \, dx = 0 = \lambda_m^4 \int \phi_m \phi_n \, dx = 0 \qquad (10.14)$$

$$\int_0^1 \phi_n'''' \phi_m \, dx = 0 = \lambda_n^4 \int \phi_m \phi_n \, dx = 0 \qquad (10.15)$$

Integrating the left side twice by parts, we get

$$\int_0^1 \phi_m'''' \phi_n \, dx = \phi_m''' \phi_n \Big|_0^1 - \phi_m'' \phi_n' \Big|_0^1 + \int_0^1 \phi_m'' \phi_n'' \, dx \qquad (10.16)$$

$$\int_0^1 \phi_m \phi_n'''' \, dx = \phi_n''' \phi_m \Big|_0^1 - \phi_n'' \phi_m' \Big|_0^1 + \int_0^1 \phi_m'' \phi_n'' \, dx \qquad (10.17)$$

Since boundary conditions are satisfied by the eigenfunctions, we have

$$\int \phi_m'' \phi_n \Big|_0^1 = 0, \int \phi_m'' \phi_n' \Big|_0^1 = 0$$

$$\int \phi_m'''' \phi_n \, dx = \int \phi_m'' \phi_n'' \, dx = \int \phi_n'''' \phi_m \, dx$$

Hence, we finally get the following equation by subtracting Equation 10.15 from Equation 10.14:

$$\left(\lambda_m^4 - \lambda_n^4\right) \int_0^1 \phi_m \phi_n \, dx = 0$$

Hence,

$$\left.\begin{array}{ll} \int_0^1 \phi_m \phi_n \, dx = \alpha_n & \text{if } m = n \\[3mm] \qquad\quad\; = 0 & \text{if } m \neq n \end{array}\right\} \phi_m \text{ and } \phi_n \text{ are orthogonal.}$$

### Response to Sinusoidal Excitation

The equation of motion is

$$EIw^{IV} + m\ddot{w} = L(y,t) \tag{10.18}$$

Let $L(y,t) = L(y) \, e^{i\omega t}$.
Let the steady-state solution be

$$w(y,t) = W(y) \, e^{i\omega t}$$

$$EIW^{IV} - m\,\omega^2 \, W = L(y)$$

Expressing

$$W(y) = \sum_{i=1}^{N} a_i \phi_i(y)$$

and

$$L(y) = \sum_{i=1}^{N} b_i \phi_i(y)$$

where

$$b_i = \frac{1}{\int \phi_i^2 \, dx} \cdot \int_0^l L(y) \phi_i(y) \, dy$$

then,

$$\sum_i EI \sum a_i \phi_i^{IV} - m\omega^2 \sum_i a_i \phi_i = \sum_i b_i \phi_i$$

From the free vibration relation,

$$EI\phi_i^{IV} = m\omega_i^2 \phi_i$$

Substituting in the equation of motion, we get

$$\sum_i a_i m\omega_i^2 \phi_i - m\omega^2 \sum_i a_i \phi_i = \sum_i b_i \phi_i$$

Multiplying throughout by $\phi_j$ and integrating over the length of the beam,

$$a_i m \int \phi_i^2 \, dx \left( \omega_i^2 - \omega^2 \right) = b_i \int \phi_i^2 \, dx$$

and hence

$$a_i = \frac{b_i}{m \left( \omega_i^2 - \omega^2 \right)} \tag{10.19}$$

## Response to Periodic Excitation

Response to periodic excitation can be obtained by following the same procedure as in the case of sinusoidal excitation. Periodic force is expressed in the form of a Fourier series and the responses to each of the terms are summed together.

## Nonairfoil Physical Problems

### Fluid Flow through a Flexible Pipe

See Figure 10.11.

Consider a hollow circular pipe of wall thickness, h; mass per unit length, $m_p = \rho_p \cdot 2\pi R \cdot h$; radius, R; density, $\rho_p$; and pipe bending stiffness, EI. The aerodynamic loading per unit length along the pipe is

$$-L = \rho A \left( \frac{\partial}{\partial t} + U \frac{\partial}{\partial x} \right)^2 w = \rho A \left( \frac{\partial^2 w}{\partial t^2} + 2U \frac{\partial^2 w}{\partial x \partial t} + U^2 \frac{\partial^2 w}{\partial x^2} \right)$$

where
  $A = \pi R^2$
  $\rho$ = fluid density
  $U$ = axial velocity
  $w$ = transverse deflection of pipe

The equation of motion of the slender pipe is given by

$$EI \frac{\partial^2 w}{\partial x^4} + m_p \frac{\partial^2 w}{\partial t^2} = L$$

This equation can also be derived from energy considerations as follows.

The kinetic energy of the fluid and the pipe and the potential energy of the pipe are, respectively, given by

$$T = \frac{1}{2} m_p \int \dot{w}^2 \, dx + \frac{1}{2} \rho A \int_0^1 \left( \frac{\partial w}{\partial t} + \frac{\partial w}{\partial x} \cdot \frac{dx}{dt} \right)^2 dx$$

$$= \frac{1}{2} m_p \int \dot{w}^2 \, dx + \frac{1}{2} \rho A \int \left( \dot{w} + \dot{w} \cdot U \right)^2 dx$$

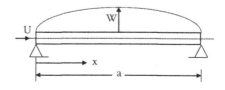

**FIGURE 10.11**
Fluid flow through flexible pipe.

$$U = \frac{1}{2}EI \int w''^2 \, dx$$

Using Euler-Lagrange equations,

$$\frac{d}{dt}\left(\frac{\partial T}{\partial \dot{w}}\right) - \frac{d}{dx}\left(\frac{\partial U}{\partial w'}\right) + \frac{d^2}{dx^2}\left(\frac{\partial U}{\partial w''}\right) = 0$$

$$m_p \ddot{w} + \rho A \ddot{w} + 2\rho AU \dot{w}' + \rho AU^2 \frac{\partial^2 w}{\partial x^2} + EI \frac{\partial^4 w}{\partial x^4} = 0$$

---

## Divergence

Let $w = 0$, $EI \dfrac{\partial^2 w}{\partial x^2} = 0$ at $x = 0$ and a. For the static case, when the deflection, w, is independent of time, we have

$$EI \frac{\partial^4 w}{\partial x^4} + \rho AU^2 \frac{\partial^2 w}{\partial x^2} = 0$$

This is similar to the buckling equation with $\rho = \rho AU^2$!
   Let $w = Ae^{px}$

$$EIp^4 + \rho AU^2 p^2 = 0$$

$$p_{1,2} = 0 \text{ and } p_{3,4} = \pm i\left(\frac{\rho AU^2}{EI}\right)^{1/2}$$

Let

$$\left(\frac{\rho AU^2}{EI}\right)a^2 = \lambda^2$$

Hence,

$$w = A_1 + A_2 x + A_3 \sin\frac{\lambda x}{a} + A_4 \cos\frac{\lambda x}{a}$$

Using boundary conditions

$$A_1 = A_2 = A_4 = 0$$

hence,

$$A_3 \sin\lambda = 0$$

that is, $\sin\lambda = 0$. Hence, $\lambda = \pi, 2\pi, 3\pi, \ldots n\pi$.
    Divergence corresponds to the lowest root, $\lambda = \pi$.
    Corresponding divergence or buckling pressure is

$$\rho U^2 = \frac{EI}{Aa^2}\pi^2$$

$\lambda^2$ is the nondimensional ratio of aerodynamic to elastic stiffness and can be termed as "aerodynamic stiffness number."
    Similarly, the terms

$$\frac{qS\dfrac{\partial C_L}{\partial\alpha}\cdot e}{K_\alpha},$$

for a typical section, and

$$\frac{q(lc)e\dfrac{\partial C_L}{\partial\alpha}}{(GJ/l)},$$

for the beam-rod wing model, are aeroelastic stiffness numbers.

### Fluid Flow over a Flexible Wall

See Figure 10.12.
    Physical situations are in nuclear reactor heat exchangers, and local skin deformations in aircraft and missiles. For one-dimensional (beam) structural representation of the wall, the equation of equilibrium is

$$EI\frac{\partial^4 w}{\partial x^4} = L$$

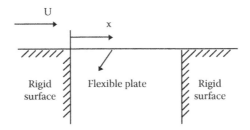

**FIGURE 10.12**
Flutter of a flexible plate with fluid flow.

Following the discussion on the fluid flow in pipes, the aerodynamic loading may be written as

$$L \approx \rho U^2 \frac{\partial^2 w}{\partial x^2}$$

and the results are the same as those for the fluid flow in flexible pipes.

## Flutter of Flexible Pipes with Fluid Flow

Physical situations are oil pipelines, hydraulic lines, rocket propellant fuel lines, and human lung airways.
    The equation of motion is

$$EI \frac{\partial^4 w}{\partial x^4} + m \frac{\partial^2 w}{\partial t^2} + \rho A \left( \frac{\partial^2 w}{\partial t^2} + 2U \frac{\partial^2 w}{\partial x \partial t} + U^2 \frac{\partial^2 w}{\partial x^2} \right) = 0$$

Consider a cantilever pipe (Figure 10.13). Let

$$w = \bar{w}(x)e^{ikt}$$

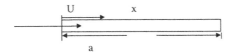

**FIGURE 10.13**
Flutter of flexible pipe with fluid flow.

Substituting the preceding in the differential equation, we get

$$\left\{ EI\frac{\partial^4 \overline{w}}{\partial x^4} - mk^2\overline{w} + \rho A\left(-k^2\overline{w} + 2Uik\frac{d\overline{w}}{dx} + U^2\frac{\partial^2\overline{w}}{\partial x^2}\right) \right\} e^{ikt} = 0$$

Assume the solution of the ordinary differential equation in the form

$$\overline{w}(x) = Ce^{Px}$$

Substituting, we get

$$EI\, p^4 - mk^2 + \rho A(-k^2 + 2Uikp + U^2p^2) = 0$$

If the four roots of the equation are $p_1$, $p_2$, $p_3$, and $p_4$,

$$\overline{w}(x) = \sum_{i=1}^{4} C_i e^{P_i x}$$

Using the following four boundary conditions,

$$\overline{w}(0) = \overline{w}'(0) = \overline{w}''(a) = \overline{w}'''(a) = 0,$$

we get

$$C_1 + C_2 + C_3 + C_4 = 0$$

$$C_1\, p_1 + C_2\, p_2 + C_3\, p_3 + C_4\, p_4 = 0$$

$$C_1 p_1^2 e^{P_1 a} + C_2 p_2^2 e^{P_2 a} + C_3 p_3^2 e^{P_3 a} + C_4 p_4^2 e^{P_4 a} = 0$$

$$C_1 p_1^3 e^{P_1 a} + C_2 p_2^3 e^{P_2 a} + C_3 p_3^3 e^{P_3 a} + C_4 p_4^3 e^{P_4 a} = 0$$

For the nontrivial solution, setting the determinant of coefficients to zero gives

$$D = \begin{vmatrix} 1 & 1 & 1 & 1 \\ p_1 & p_2 & p_3 & p_4 \\ p_1^2 e^{P_1 a} & p_2^2 e^{P_2 a} & p_3^2 e^{P_3 a} & p_4^2 e^{P_4 a} \\ p_1^3 e^{P_1 a} & p_2^3 e^{P_2 a} & p_3^3 e^{P_3 a} & p_4^3 e^{P_4 a} \end{vmatrix} = 0$$

There is no analytical solution for this problem. Numerical solutions are obtained by giving a guess value of k for a given pipe with given U.

$p_1$, $p_2$, $p_3$, and $p_4$ are then evaluated. In general, corresponding D will not be zero unless k is right. For U = 0, k is imaginary and so are the natural frequencies. For U > 0, k is complex. Using nondimensional terms as follows,

$$\beta = \rho A/(\rho A + m)$$

$$u = \left( \rho \frac{AU^2}{EI} \right)^{\frac{1}{2}} \cdot a$$

$$\Omega = [(m + \rho A)/EI]^{1/2} \cdot ka^2$$

When the real part of $k_1$ becomes positive, flutter occurs (Figure 10.14).

The stability boundary for this system may be represented in terms of u and $\beta$ as seen in Figure 10.15. Also shown is frequency $\Omega_F$, of the flutter oscillation.

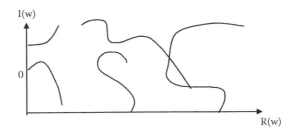

**FIGURE 10.14**
Conditions for flutter.

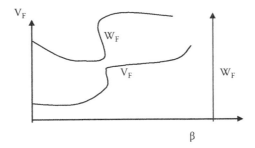

**FIGURE 10.15**
Stability boundaries.

## (High Speed) Fluid Flow over a Flexible Wall-Plate Flutter

This becomes important with the advent of supersonic flight. The problem is shown in Figure 10.16.

Consider a simplified model of three rigid plates, each hinged at both ends. The middle two hinges are supported by springs as shown in Figure 10.17.

Plate mass per unit length is m and length is l. When $M \gg 1$ (Mach number), the aerodynamic pressure change, p, due to plate motion is

$$p = \frac{p_\infty U_\infty^2 \partial w}{M_\infty \partial x}$$

There are two degrees of freedom; choose $q_1$ and $q_2$ as the deflection of the two springs. The potential energy in the springs is

$$U = \frac{1}{2} kq_1^2 + \frac{1}{2} kq_2^2$$

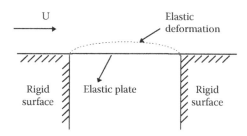

**FIGURE 10.16**
Flutter of a flexible plate with fluid flow.

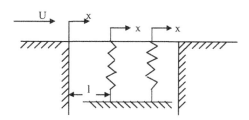

**FIGURE 10.17**
Fluid flow over spring-supported plate.

We then have

Plate 1: $w = q_1 \dfrac{x}{1}$; $\dfrac{\partial w}{\partial x} = \dfrac{q_1}{1}$

Plate 2: $w = q_1 \left(1 - \dfrac{x}{1}\right) + q_2 \dfrac{x}{1}$; $\dfrac{\partial w}{\partial x} = \dfrac{q_2 - q_1}{1}$

Plate 3: $w = q_2 \left(1 - \dfrac{x}{1}\right)$; $\dfrac{\partial w}{\partial x} = \dfrac{-q_2}{1}$

The kinetic energy is given by

$$T = \frac{1}{2} \int m \left(\frac{\partial w}{\partial t}\right)^2 dx$$

$$= \frac{1}{2} ml \left[\frac{2}{3} \dot{q}_1^2 + \frac{2}{3} \dot{q}_2^2 + \frac{2}{6} \dot{q}_1 \dot{q}_2\right]$$

Virtual work by aerodynamic pressure is given by

$$\delta W = \int (-p)\, \delta w\, dx \qquad p = \frac{\rho_\infty U_\infty^2}{M_\infty} \cdot \frac{\partial w}{\partial x}$$

$$= Q_1\, \delta q_1 + Q_2\, \delta q_2 \qquad = A \frac{\partial w}{\partial x}$$

where

$$Q_1 = -\frac{\rho_\infty U_\infty^2}{M_\infty} \cdot \frac{q_2}{2}$$

$$Q_2 = \frac{\rho_\infty U_\infty^2}{M_\infty} \cdot \frac{q_1}{2}$$

Using Lagrange's equations,

$$\frac{2}{3} ml\ddot{q}_1 + \frac{ml}{2} \ddot{q}_2 + kq_1 + \frac{\rho_\infty U_\infty^2}{2M_\infty} \cdot q_2 = 0$$

$$\frac{ml}{6}\ddot{q}_1 + \frac{2}{3}ml\ddot{q}_2 + kq_2 - \frac{\rho_\infty U_\infty^2}{2M_\infty} \cdot q_1 = 0$$

Let

$$q_1 = \bar{q}_1 e^{i\omega t},$$

$$q_2 = \bar{q}_2 e^{i\omega t}$$

Then,

$$\left[ -\omega^2 ml \begin{bmatrix} 2/3 & 1/6 \\ 1/6 & 2/3 \end{bmatrix} + \begin{bmatrix} k & 0 \\ 0 & k \end{bmatrix} + \frac{\rho_\infty U_\infty^2}{2M_\infty} \begin{bmatrix} 0 & 1 \\ -1 & 1 \end{bmatrix} \right] \begin{Bmatrix} q_1 e^{i\omega t} \\ q_2 e^{i\omega t} \end{Bmatrix} = \{0\}$$

For a nontrivial solution, we have the determinant of coefficients equated to zero. Hence,

$$\frac{15}{36}\Omega^4 - \frac{4}{3}\Omega^2 + 1 + \lambda^2 = 0$$

where

$$\Omega^2 = \frac{\omega^2 ml}{k},$$

and

$$\lambda = \frac{\rho \rho_\infty U_\infty^2}{2M_\infty k}$$

Then we have

$$\Omega^2 = \frac{8}{5} \pm \frac{2}{5}(1 - 15\lambda^2)^{1/2}$$

When the argument of the square root is negative, $\Omega^2$ becomes complex, and hence one solution for $\Omega$ will have a negative imaginary part corresponding to an unstable solution. Hence, flutter will occur for

$$\lambda^2 > \lambda_F^2 = \frac{1}{15}$$

The frequency at this $\lambda_F$ is

$$\Omega_F = \left(\frac{8}{5}\right)^{1/2}$$

For reference, the natural frequencies ($\lambda \equiv 0$) are

$$\Omega_1 = \left(\frac{6}{5}\right)^{1/2} \text{ and } \Omega_2 = (2)^{1/2}$$

Further, from the equations, we get the following:

The first normal mode: $\dfrac{\bar{q}_1}{\bar{q}_2} = +1$ for $\Omega = \Omega_1$ at $\lambda = 0$.

The second natural mode: $\dfrac{\bar{q}_1}{\bar{q}_2} = -1$ for $\Omega = \Omega_2$ at $\lambda = 0$.

The flutter mode: $\dfrac{\bar{q}_1}{\bar{q}_2} = -4 + 15^{1/2}$ for $\Omega = \Omega_F$, $\lambda = \lambda_F$.

Sketches of the corresponding plate shapes are given in Figure 10.18. Note that

1. The flutter mechanism is a convergence of natural frequencies with increasing flow velocities. Flutter frequency is between the first and second frequencies.
2. The flutter mode shape shows a maximum near the rear edge of the plate.

This rigid plate model is an oversimplification. There is no curvature effect here.

1st mode:

2nd mode:

Flutter mode:

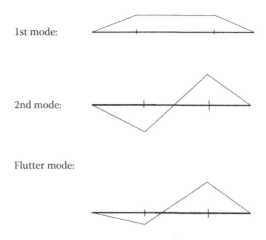

**FIGURE 10.18**
Flutter modes.

## Stall Flutter

Stall flutter is a phenomenon that occurs with partial or complete breakaway of the flow from the airfoil periodically, during oscillation. Unlike in classical flutter (i.e., when the flow remains attached at all times), the mechanism for energy transfer from the air stream to the oscillating airfoil relies on neither elastic nor aerodynamic coupling between two modes or upon a phase lag between a displacement and its aerodynamic reaction. An essential feature of stall flutter is the nonlinear aerodynamic reaction to the motion of the airfoil.

Stall flutter of aircraft wings is associated with very high angles of attack.

Electric power transmission cables of circular cross sections modified by bundling or ice accretion are prone to stall flutter, which is called "galloping." Suspension bridges, helicopter rotors, and turbomachinery blades are also prone to stall flutter.

In a series of tests, the field of flow around an airfoil oscillating about the stalling angle was studied. In a stroke of increasing amplitude, the separation is delayed to an angle of attack appreciably greater than that for a stationery airfoil. On the return movement, reestablishment of a smooth flow is also delayed. This phenomenon was called "aerodynamic hysteresis," which is the basic cause of stall flutter.

Stall flutter can be delayed if the airfoil can be prevented from stalling.

Divergence speed of a propeller is an important parameter to consider with stall flutter. If the rotational speed of a propeller is so high that the relative wind speed is close to the critical divergence speed, the blade will be twisted excessively, possibly beyond the stalling angle, causing stall flutter.

A separated flow over the wing creates aerodynamic forces that must be regarded as a stochastic process. The resulting flutter is called buffeting flutter.

## Flutter and Buffeting of Bridges

Let h and $\alpha$ be the vertical and torsional degrees of freedom of a bridge deck section model, referred to as its elastic axis (Figure 10.19).

$$m\ddot{h} + S_\alpha\ddot{\alpha} + c_h\dot{h} + K_h h = L_h$$

$$S_\alpha\ddot{h} + I\ddot{\alpha} + c_\alpha\dot{\alpha} + K_\alpha\alpha = M_\alpha$$

where $c_h$ and $c_\alpha$ are damping coefficients. $L_h$ and $M_\alpha$ are aerodynamic lift and moment at the elastic axis.

The majority of bridge sections on bridge decks are symmetric about their centerlines and have $S_\alpha \equiv 0$

The basic models of self-excited (flutter) forces and moments on bridge decks are

$$L_h = m[H_1\dot{h} + H_2\dot{\alpha} + H_3\alpha]$$

$$M_\alpha = I[A_1\dot{h} + A_2\dot{\alpha} + A_3\alpha]$$

with $H_i$, $A_i$ (i = 1, 2, 3) to be determined experimentally. In the context of heavy civil engineering structures, aerodynamic inertial terms in $\dot{h}$, $\ddot{\alpha}$ are usually considered negligible.

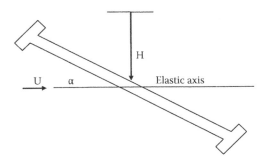

**FIGURE 10.19**
Bridge buffeting.

Buffeting is an irregular motion of a structure or parts of a structure in a flow, excited by turbulence in the flow.

---

## Aeroelasticity of Turbomachinery Blades

Divergence is not a significant problem in turbomachines. Due to unsteadiness of the aerodynamic or external forces, the blades assume a series of time-dependent positions. A further complication is that a blade cannot be considered as an isolated structure. There exists aerodynamic and structural coupling between neighboring blades, which dictate a model description of the entire bladed disk assembly.

### Some Examples

1. During normal operation of nuclear steam generators, the situation may occur when one or more tubes are degraded by wear, fretting, or forms of chemical attack. Once a flaw is initiated, usually at the secondary face of the tubesheets or near the tube support plates, flow-induced vibration may accelerate the degradation.

2. With rapid developments in power, nuclear, and other industries, good overall performance of heat exchangers is required. The appearance of large and high-performance heat exchangers has brought problems with flow-induced vibrations due to (1) vortex shedding, (2) turbulent buffeting, and (3) fluid elastic instability.

3. Intense acoustic resonance results from gas flow across a bank of heat exchanger tubes. Sound levels as high as 173 dB have been observed within the bundle. During resonance, the sound correlates vortex shedding from the tubes.

4. The vibration of propellers of high-speed ships is called propeller-excited vibration and is caused by the oscillatory forces set up as the propeller rotates in an uneven wake flow.

# 11

## Nonlinear Aeroelasticity

### Generic Nonlinear Aeroelastic Behavior

In the aeroelasticity literature, a linear system may mean either a system whose static and dynamic responses are linear or one that is nonlinear in its static response, but linear in its dynamic response. Therefore, the term "linear system" must be qualified by specifying whether the system is dynamically linear or both statically and dynamically (wholly) linear.

A system that is wholly linear is one whose deformation to either static or dynamic forces is proportional to those forces. An aerodynamic flow is wholly linear when the response, such as a change in pressure, is proportional to changes in downwash or fluid velocities induced by geometry or motion of a rigid body in the flow.

A statically nonlinear but dynamically linear structure is one whose static deformations are sufficiently large to enter into the nonlinear region. In aerodynamic flows, shockwaves and separated flows are themselves the result of dynamically nonlinear processes. However, once formed, they may often be treated as part of a nonlinear static equilibrium state or steady flow.

### One Degree of Freedom Nonlinear System: Divergence

Consider the typical section model for the aircraft wing discussed in Chapter 4, which is shown again in Figure 11.1.

The angle of attack is given by

$$\alpha = \alpha_o + \alpha_e \tag{11.1}$$

where $\alpha_o$ = initial angle of attack, which is assumed to be zero in the ensuing discussion.

$$\alpha_e = \text{elastic twist of spring}$$

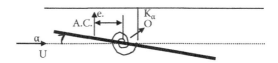

**FIGURE 11.1**
A typical section of airfoil.

Moment about the elastic axis is

$$M_y = M_{AC} + L \cdot e \tag{11.2}$$

where
  $M_y$ = moment about elastic axis
  $M_{AC}$ = moment about aerodynamic center
  $L$ = lift, net vertical force, positive up
  $e$ = distance from elastic axis to aerodynamic center, positive toward nose

From aerodynamic theory,

$$L = C_L \cdot qS \tag{11.3}$$

$$M_{AC} = C_{MAC} \cdot qSc \tag{11.4}$$

where S is the wing area and c is the mean chord length. Assuming that the aerodynamic moment is negligible, we have $C_{MAC} = 0$. Lift coefficient varies with $\alpha$ and, using Taylor series expansion,

$$C_L = \frac{\partial C_L}{\partial \alpha} \cdot \alpha \tag{11.5}$$

Consider the linear case with a linear spring of constant $K_\alpha$, with the elastic moment of $-K_\alpha \cdot \alpha_e$. Hence, for equilibrium,

$$M_y - K_\alpha \cdot \alpha_e = 0$$

Substituting for $M_y$, we get

$$\left(\frac{\partial C_L}{\partial \alpha} \cdot \alpha\right) \cdot qSe - K_\alpha \cdot \alpha_e = 0 \tag{11.6}$$

Rewriting in the form of

$$\alpha_e \left( K_\alpha - \frac{\partial C_L}{\partial \alpha} \cdot qS \cdot e \right) = 0 \tag{11.7}$$

we have either $\alpha = 0$ or

$$1 - \frac{qSe}{K_\alpha} \cdot \frac{\partial C_L}{\partial \alpha} = 0 \tag{11.8}$$

Under these conditions, the divergence dynamic pressure q is given by

$$q_D = \frac{K_\alpha}{S \cdot e \cdot \dfrac{\partial C_L}{\partial \alpha}} \tag{11.9}$$

Since $q = \frac{1}{2}\rho U^2$, the divergence speed is given by

$$U_D = \sqrt{\frac{2q_D}{\rho}} \tag{11.10}$$

## Nonlinear Spring

Let us assume that the elastic moment for the nonlinear spring is given by

$$M_E = -K_\alpha \cdot \alpha_e - K_{\alpha 3} \cdot \alpha_e^3 \tag{11.11}$$

with $K_\alpha > 0$ and $K_{\alpha 3} > 0$. Also, expanding lift in the Taylor series, we get

$$L = qs[(\partial C_L / \partial \alpha) \cdot \alpha - (\partial C_L / \partial \alpha)_3 \cdot \alpha^3] \tag{11.12}$$

where $\left( \dfrac{\partial C_L}{\partial \alpha} \right)_3 = \dfrac{1}{3!} \cdot \dfrac{\partial^3 C_L}{\partial \alpha^3}$. Note that lift decreases for large $\alpha$ due to the flow separation.

Assuming $\alpha_o = C_{MAC} = 0$, we have

$$eqS \left[ \frac{\partial C_L}{\partial \alpha} \cdot \alpha_e - \left( \frac{\partial C_L}{\partial \alpha} \right)_3 \alpha_e^3 \right] - \left[ K_\alpha \cdot \alpha_e + K_{\alpha 3} \cdot \alpha_e^3 \right] = 0 \tag{11.13}$$

Rearranging,

$$\alpha_e \left[ eqS\frac{\partial C_L}{\partial \alpha} - K_\alpha \right] - \alpha_e^3 \left[ eqS \left( \frac{\partial C_L}{\partial \alpha} \right)_3 + K_{\alpha 3} \right] = 0 \qquad (11.14)$$

Solving, we have $\alpha_e = 0$ or

$$\alpha_e^2 = \frac{\left[ eqS\dfrac{\partial C_L}{\partial \alpha} - K_\alpha \right]}{\left[ eqS \left( \dfrac{\partial C_L}{\partial \alpha} \right)_3 + K_{\alpha 3} \right]} \qquad (11.15)$$

$$= \frac{\dfrac{\partial C_L}{\partial \alpha}}{\left( \dfrac{\partial C_L}{\partial \alpha} \right)_3} \frac{(q - q_D)}{(q + q_{D3})}$$

where

$$q_D = \frac{K_\alpha}{Se\dfrac{\partial C_L}{\partial \alpha}} \quad \text{and} \quad q_{D3} = \frac{-K_{\alpha 3}}{Se \left( \dfrac{\partial C_L}{\partial \alpha} \right)_3}.$$

The right-hand side in Equation 11.15 must be positive for $\alpha_e$ to be real. For simplicity, assume $e > 0$. Then nontrivial solutions are possible only for $q > q_D$. This is illustrated in Figure 11.2. It should be remembered that $q_D$ corresponds to the linear case and has no physical significance for the nonlinear case.

For $q < q_D$, only the first root $\alpha_e = 0$ is possible. Hence, for $q < q_D$, static equilibrium is not possible.

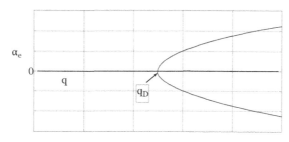

**FIGURE 11.2**
Equation for elastic twist $e > 0$.

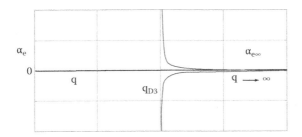

**FIGURE 11.3**
Equation for elastic twist e < 0.

Note that two symmetrical equilibrium positions are possible for $q > q_D$. In this case, $\alpha_e$ remains finite for any finite $q > q_D$. This situation is illustrated in Figure 11.3.

For $e < 0$, we have (from denominator of Equation 11.15)

$$q_{D3} = K_{\alpha3}/eS(\partial C_L/\partial\alpha)_3 \tag{11.16}$$

and

$$\alpha_{e\infty}^2 = \frac{\partial C_L/\partial\alpha}{(\partial C_L/\partial\alpha)_3}$$

(by dividing numerator and denominator by $q \to \infty$).

## One Degree of Freedom System: Dynamic Effects

Consider the $\alpha$ degree of freedom motion. We can write the equation of motion in the form of a Duffing's equation as

$$\alpha'' + c\alpha' + k\alpha + \beta\alpha^3 = F\sin(\omega\tau) \tag{11.17}$$

We will neglect the damping term $c\alpha'$ in the following discussion. The relation between the amplitude, $R$, and the frequency, $\omega$, for the preceding nonlinear equation is given by

$$\omega^2 = k + \frac{3}{4}\beta R^2 - \frac{F}{R} \tag{11.18}$$

For certain values of the parameters, Equation 11.18 gives three values of R for one value of $\omega$. This leads to a jump phenomenon where the solution "jumps" from one branch of the amplitude–frequency curve to another.

It should be noted that these approximations do not yield any information about the effect of initial conditions on the final solution after the transients have died down. In other words, the effect of the initial conditions that determines the branch of the jump where the solution lies is not clear from the analytical studies. We can assess the effect of the initial conditions only numerically.

## Two Degrees of Freedom System: Dynamic Effects

Consider a two degrees of freedom dynamic system with inertia coupling. The coefficients are arbitrarily chosen to generate a set of differential equations. It is assumed that the response is harmonic and the amplitudes are slowly varying functions of time. The example considers the two uncoupled natural frequencies to be equal.

The equations of motion are given as

$$x_a\alpha'' + \xi'' + \xi + 0.25\xi^3 = 0$$
$$\alpha'' + x_a\xi'' + \alpha + 0.25\alpha^3 = F\sin(\omega\tau) \tag{11.19}$$

The coefficients of these two coupled equations are similar.

In order to see the response characteristics for a one degree of freedom system with these system parameters, let us decouple the equations into single degree of freedom systems by taking $x_a = 0$. This is similar to the uncoupled case in Equation 11.17. The coupled system is governed by Equation 11.19.

# 12

## Unsteady Aerodynamics

### Introduction

This chapter on unsteady aerodynamics is included only to emphasize the fact that the angle of attack changes not only with the twist angle $\theta$ but also on the flow conditions. In flutter analysis, using an aerodynamic theory for steady flow considers lift and pitching moment as functions of the instantaneous pitch angle, $\theta$, only. However, the angle of attack is not simply equal to $\theta$. Consider the airfoil reference point plunging with velocity $\dot{h}$, at least for small angles. The angle of attack including the effect of plunge is given by

$$\alpha = \theta + \frac{\dot{h}}{U} \tag{12.1}$$

Consequently, unsteady-aerodynamic theories need to account for at least three separate physical phenomena, as follows:

1. In view of the airfoil's unsteady motion relative to the air, the relative wind vector is not fixed in space. The changing direction of the relative wind changes the effective angle of attack and thus changes the lift.

2. The airfoil motion disturbs the flow and causes a vortex to be shed at the trailing edge (Fung 1955). The downwash from this vortex, in turn, changes the flow that impinges on the airfoil. This unsteady downwash changes the effective angle of attack and thus changes the lift.

3. The motion of the airfoil accelerates air particles near the airfoil surface, thus creating the need to account for the resulting inertial forces (although this "apparent-inertia" effect is less significant than that of the shed vorticity). The apparent-inertia effect does not change the angle of attack but it does, in general, affect both lift and pitching moment.

---

## Theodorsen's Unsteady Thin-Airfoil Theory

Theodorsen derived a theory of unsteady aerodynamics for a thin airfoil performing small, simple harmonic oscillations in incompressible flow. He assumed linear potential-flow theory. The lift includes both circulatory and noncirculatory terms. However, the pitching moment about the quarter-chord is completely noncirculatory. Theodorsen's theory gives the lift and pitching moment as

$$L = 2\pi\rho_\infty UbC(k)\left[\dot{h} + U\theta + b\left(\frac{1}{2} - a\right)\dot{\theta}\right] + \pi\rho_\infty b^2(\ddot{h} + U\dot{\theta} - ba\ddot{\theta})$$

$$M_{\frac{1}{4}} = -\pi\rho_\infty b^3\left[\frac{1}{2}\ddot{h} + U\dot{\theta} + b\left(\frac{1}{8} - \frac{a}{2}\right)\ddot{\theta}\right]$$

(12.2)

The function $C(k)$, known as Theodorsen's function, is complex and is a function of the reduced frequency, $k$, given by

$$C(k) = \frac{H_1^{(2)}(k)}{H_1^{(2)}(k) + iH_0^{(2)}(k)}$$

(12.3)

The functions $H_n^{(2)}(k)$ are Hankel functions of the second kind, expressed in terms of Bessel functions of the first and second kind, as

$$H_n^{(2)}(k) = J_n(k) - iY_n(k)$$

(12.4)

In Theodorsen's theory, the lift-curve slope is equal to $2\pi$. The first of the two terms in the lift is the circulatory lift without the effect of shed vortices multiplied by $C(k)$, which is because the theory considers the effect of shed vorticity. The noncirculatory terms in the second term in the lift, as well as the entire pitching-moment expression, depend on the acceleration and angular acceleration of the airfoil. These are mostly apparent-mass/apparent-inertia terms. The coefficient of $\ddot{h}$ in the lift is the mass per unit length of the air contained in an infinitely long circular cylinder of radius $b$. This indicates how much air is given the acceleration by motion of the airfoil.

For steady flow, the circulatory lift is linear in terms of the angle of attack. But for unsteady flow, the flow direction varies along the chord line as the result of the induced flow varying along the chord, and hence there is no single angle of attack. However, for unsteady flow, an effective angle of attack will be useful. For simple harmonic motion, Theodorsen's theory leads to an effective angle of attack given by

$$\alpha = C(k)\left[\theta + \frac{\dot{h}}{U} + \frac{b}{U}\left(\frac{1}{2} - a\right)\dot{\theta}\right]$$ (12.5)

It must be emphasized that Theodorsen's theory with the C(k) is valid only for simple harmonic motion.

Theodorsen's theory with C(k) taken as unity is called a "quasi-steady" thin-airfoil theory. Such an approximation is valid only for cases where k is very small. For slow harmonic oscillations or slowly varying motions which are not harmonic, the quasi-steady theory may be used in the time domain.

## Flutter Prediction via Assumed Modes

Aircraft structures are complex, with elements of different geometry and structural properties. The finite element method can realistically represent dynamics of aircraft structures. However, a full finite-element flutter analysis will involve extremely large elements and degrees of freedom. Flutter analysis based on a truncated set of modes of the structure is still desirable. It will be approximate, however, and it will be relatively simple. In the following, the Rayleigh-Ritz method is employed considering the aircraft wing as a uniform cantilever beam of length l. The strain energy for the wing is given by

$$U = \frac{1}{2}\int_0^l \left[EI\left(\frac{\partial^2 w}{\partial y^2}\right)^2 + GJ\left(\frac{\partial \theta}{\partial y}\right)^2\right]dy$$ (12.6)

with bending rigidity EI and torsional rigidity GJ.

The kinetic energy is given as

$$K = \frac{1}{2}\int_0^l \iint_A \rho\left[\left(\frac{\partial w}{\partial t} - x\frac{\partial \theta}{\partial t}\right)^2 + z^2\left(\frac{\partial \theta}{\partial t}\right)^2\right]dxdydz$$ (12.7)

where the mass per unit volume of the material is ρ and the velocity of a typical point within the cross-sectional plane is given by

$$v = z\frac{\partial \theta}{\partial t}\hat{i} + \left(\frac{\partial w}{\partial t} - x\frac{\partial \theta}{\partial t}\right)\hat{k}$$ (12.8)

Here, $\hat{i}$ and $\hat{k}$ are unit vectors in the x and z directions, respectively. Accordingly, the kinetic energy can be expressed as

$$K = \frac{1}{2}\int_0^1\left[m\left(\frac{\partial w}{\partial t}\right)^2 + 2md\frac{\partial w}{\partial t}\frac{\partial \theta}{\partial t} + mb^2r^2\left(\frac{\partial \theta}{\partial t}\right)^2\right]dy \qquad (12.9)$$

where
  m is the mass per unit length
  d is the offset of the mass centroid from the elastic axis (considered positive when the mass centroid is toward the leading edge)
  b is the semichord
  br is the cross-sectional mass radius of gyration about the elastic axis

The virtual work of the aerodynamic forces can be written as

$$\overline{\delta W} = \int_0^1\left[L'\delta w + \left(M'_{ac} + eL'\right)\delta\theta\right]dy \qquad (12.10)$$

where $L'$ and $M'_{ac}$ are the distributed lift and pitching moment per unit length of the wing.

Rewriting the strain energy, kinetic energy, and virtual work using the following notations,

$$d \to -bx_\theta$$

$$e \to \left(\frac{1}{2}+a\right)b$$

$$L' \to L'$$

$$M'_{ac} = M'_{\frac{1}{4}}$$

the strain energy is the same as before. The kinetic energy is given by

$$K = \frac{1}{2}\int_0^1\left[m\left(\frac{\partial w}{\partial t}\right)^2 - 2mbx_\theta\frac{\partial w}{\partial t}\frac{\partial \theta}{\partial t} + mb^2r^2\left(\frac{\partial \theta}{\partial t}\right)^2\right]dy \qquad (12.11)$$

and the virtual work is given by

$$\overline{\delta W} = \int_0^1 \left\{ L'\delta w + \left[ M_1' + \left( \frac{1}{2} + a \right) bL' \right] \delta\theta \right\} dy \qquad (12.12)$$

The assumed modes are the set of free-vibration modes in bending and torsion for the beam and are expressed as

$$w(y,t) = \sum_{i=1}^{N_w} \eta_i(t)\Psi_i(t)$$

$$\theta(y,t) = \sum_{i=1}^{N_\theta} \phi_i(t)\Theta_i(t) \qquad (12.13)$$

Here, $N_w$ and $N_\theta$ are the numbers of modes in bending and torsion, respectively; $\eta_i$ and $\phi_i$ are the generalized coordinates associated with bending and torsion, respectively; and $\Psi_i$ and $\Theta_i$ are the bending and torsion mode shapes, respectively. Further, $\Theta_i$ is given by

$$\Theta_i = \sqrt{2}\sin(\gamma_i y) \qquad (12.14)$$

where

$$\gamma_i = \frac{\pi\left( i - \frac{1}{2} \right)}{1} \qquad (12.15)$$

and $\Psi_i$ is given as (Young and Felgar 1949)

$$\Psi_i = \cosh(\alpha_i y) - \cos(\alpha_i y) - \beta_i[\sinh(\alpha_i y) - \sin(\alpha_i y)] \qquad (12.16)$$

with $\alpha_i$ and $\beta_i$ as given in Young and Felgar (1949).

Discretizing spatially the strain energy, kinetic energy, and virtual work and using the orthogonality property among the bending and torsion modes, the strain energy simplifies to

$$U = \frac{1}{2}\left[ \frac{\overline{EI}}{1^3} \sum_{i=1}^{N_w} (\alpha_i 1)^4 \eta_i^2 + \frac{\overline{GJ}}{1} \sum_{i=1}^{N_\theta} (\gamma_i 1)^2 \phi_i^2 \right] \qquad (12.17)$$

The kinetic energy is also simplified because of the orthogonality among the bending and torsion modes and can be written as

$$K = \frac{ml}{2}\left( \sum_{i=1}^{N_w} \dot{\eta}_i^2 + b^2 r^2 \sum_{i=1}^{N_\theta} \dot{\phi}_i^2 - 2bx_\theta \sum_{i=1}^{N_\theta} \sum_{j=1}^{N_w} A_{ij}\dot{\eta}_j \right) \tag{12.18}$$

where

$$A_{ij} = \frac{1}{l} \int_0^1 \Theta_i \Psi_j \, dy \; i = 1, 2, \ldots, N_\theta \; j = 1, 2, \ldots, N_w \tag{12.19}$$

There is inertial coupling between bending and torsion motion. The term involving $A_{ij}$ is a fully populated matrix because the bending and torsion modes are not orthogonal to one another. Using the virtual-work expression,

$$\overline{\delta W} = \sum_{i=1}^{N_w} \Xi_{wi}\delta\eta_i + \sum_{i=1}^{N_\theta} \Xi_{\theta i}\delta\phi_i \tag{12.20}$$

To identify the generalized forces, we have

$$\Xi_{wi} = \int_0^1 \Psi_i L' \, dy$$

$$\Xi_{\theta i} = \int_0^1 \Theta_i \left[ M'_{\frac{1}{4}} + \left( \frac{1}{2} + a \right) bL' \right] dy \tag{12.21}$$

where expressions for $L'$ and $M'_{\frac{1}{4}}$ can be found as follows. Using Theodorsen's theory, we have

$$L' = 2\pi\rho_\infty U b C(k)\left[ U\theta - \frac{\partial w}{\partial t} + b\left( \frac{1}{2} - a \right)\frac{\partial\theta}{\partial t} \right] + \pi\rho_\infty b^2 \left( U\frac{\partial\theta}{\partial t} - \frac{\partial^2 w}{\partial t^2} - ba\frac{\partial^2\theta}{\partial t^2} \right)$$

$$M'_{\frac{1}{4}} = -\pi\rho_\infty b^3 \left[ U\frac{\partial\theta}{\partial t} - \frac{\partial^2 w}{2\partial t^2} + b\left( \frac{1}{8} - \frac{a}{2} \right)\frac{\partial^2\theta}{\partial t^2} \right]$$

$$\tag{12.22}$$

The generalized forces are put easily into matrix form as

$$
\left\{ \begin{array}{c} \Xi_w \\ \Xi_\theta \end{array} \right\} = -\pi\rho_\infty b^2 l \begin{bmatrix} [\Delta] & ba[A]^T \\ ba[A] & b^2\left(a^2 + \dfrac{1}{8}\right)[\Delta] \end{bmatrix} \left\{ \begin{array}{c} \ddot{\eta} \\ \ddot{\phi} \end{array} \right\}
$$

$$
-\pi\rho_\infty bUl \begin{bmatrix} 2C(k)[\Delta] & -b\left[1+2\left(\dfrac{1}{2}-a\right)C(k)[A]^T\right] \\ 2b\left(\dfrac{1}{2}+a\right)C(k)[A] & b^2\left(\dfrac{1}{2}-a\right)\left[1-2\left(\dfrac{1}{2}+a\right)C(k)\right][\Delta] \end{bmatrix} \left\{ \begin{array}{c} \dot{\eta} \\ \dot{\phi} \end{array} \right\}
$$

$$
-\pi\rho_\infty bU^2 l \begin{bmatrix} [0] & -2C(k)[A]^T \\ [0] & -2(1+2a)C(k)[\Delta] \end{bmatrix} \left\{ \begin{array}{c} \eta \\ \phi \end{array} \right\}
$$

$$(12.23)$$

where $[\Delta]$ denotes an identity matrix and $[0]$ denotes a matrix of zeros. In view of the inherent limitations in deriving Theodorsen's theory, the generalized forces are valid only for simple harmonic motion. The rectangular submatrices in this equation are also referred to as aerodynamic-influence coefficients.

Using the condition of stationarity of the natural frequencies with respect to the normal modes provides the generalized equations of motion, which are written in the matrix form as

$$
ml \begin{bmatrix} [\Delta] & -bx_\theta[A]^T \\ -bx_\theta[A] & b^2r^2[\Delta] \end{bmatrix} \left\{ \begin{array}{c} \ddot{\eta} \\ \ddot{\phi} \end{array} \right\} + \begin{bmatrix} \dfrac{\overline{EI}}{l^3}[B] & [0] \\ [0] & \dfrac{\overline{GJ}}{l}[T] \end{bmatrix} \left\{ \begin{array}{c} \eta \\ \phi \end{array} \right\} = \left\{ \begin{array}{c} \Xi_w \\ \Xi_\theta \end{array} \right\}
$$

$$(12.24)$$

Here, the elements of the diagonal matrices $['B']$ and $['T']$ are given by

$$
B_{ii} = (\alpha_i l)^4
$$
$$
T_{ii} = (\gamma_i l)^2
$$

$$(12.25)$$

The diagonal matrices $['B']$ and $['T']$ in the stiffness matrix and the $\Delta$ in the mass matrix and generalized forces are in view of the orthogonality of the chosen basis functions $\Theta_i$ and $\Psi_i$. Although such orthogonality among the assumed modes is not a necessary condition in the application of the Rayleigh-Ritz method, such a choice simplifies the discretized equations.

Let

$$\bar{\eta}(t) = \bar{\eta}\exp(i\omega t)$$

$$\phi(t) = \bar{\phi}\exp(i\omega t)$$

(12.26)

where $\omega$ is the frequency of the simple harmonic motion.

Consider the case of $N_w = N_\theta = 1$. The equations of motion can be cast in the form of

$$\left\{ \mu\left[1 - \left(\frac{\omega_w}{\omega}\right)^2\right] + l_w \right\}\frac{\bar{\eta}_1}{b} + (l_\theta - \mu x_\theta)A_{11}\bar{\phi}_1 = 0$$

$$(m_w - \mu x_w)A_{11}\frac{\bar{\eta}_1}{b} + \left\{ \mu r^2\left[1 - \left(\frac{\omega_\theta}{\omega}\right)^2\right] + m_\theta \right\}\bar{\phi}_1 = 0$$

(12.27)

Using loads from Theodorsen's theory, $l_w$, $l_\theta$, $m_w$, and $m_\theta$ are given as

$$l_w = 1 - \frac{2iC(k)}{k}$$

$$l_\theta = a + \frac{i}{k}\left[1 + 2\left(\frac{1}{2} - a\right)C(k)\right] + \frac{2C(k)}{k^2}$$

$$m_w = a - \frac{2i\left(\frac{1}{2} - a\right)C(k)}{k}$$

(12.28)

$$m_\theta = a^2 + \frac{1}{8} - \frac{\left(\frac{1}{2} + a\right)\left[1 - 2\left(\frac{1}{2} - a\right)C(k)\right]i}{k} + \frac{2\left(\frac{1}{2} + a\right)C(k)}{k^2}$$

The fundamental bending and torsion frequencies are given by

$$\omega_w = (\alpha_1 l)^2\sqrt{\frac{EI}{ml^4}}$$

$$\omega_\theta = \frac{\pi}{2}\sqrt{\frac{GJ}{mb^2r^2l^2}}$$

(12.29)

# References

Bisplinghoff, R. L., H. Ashley, and R. L. Halfman. 1955. *Aeroelasticity*. Boston: Addison-Wesley Publishing Co., Inc.

Curtiss, H. C., Jr., R. H. Scanlan, and F. Sisto. 1989. *A modern course in aeroelasticity*, ed. E. H. Dowell. Boston: Kluwer Academic Publishers.

Fung, Y. C. 1955. *An introduction to the theory of aeroelasticity*. New York: John Wiley & Sons, Inc.

Pines, S. 1958. An Elementary Explanation of the Flutter Mechanism, *Proceedings of Dynamics and Aeroelasticity Meeting*, New York: I. A. S., pp. 52–59.

Shapiro, A. H. 1964. *Shape and flow: The fluid dynamics of drag (Science Study Series)*. London: Heinemann.

Young, D. and R. P. Felgar. 1949. *Tables of characteristic functions representing normal modes of vibration of a beam*. Austin: University of Texas in Austin.

# *Index*